PLUMBING REPAIRS SIMPLIFIED

Donald R. Brann

Library of Congress Card No. 67-27691

TENTH PRINTING — 1976
REVISED EDITION

Published by
DIRECTIONS SIMPLIFIED, INC.
Division of
EASI-BILD PATTERN CO., INC.
Briarcliff Manor, NY 10510

FIRST PRINTING

REVISED EDITIONS
1968,1969,1971,1972,
1973,1974,1975,1976

ISBN 0-87733-675-X

NOTE
Due to the variance in quality and availability of many materials and products, always follow directions a manufacturer and/or retailer offers. Unless products are used exactly as the manufacturer specifies, its warranty can be voided. While the author mentions certain products by trade name, no endorsement or end use guarantee is implied. In every case the author suggests end uses as specified by the manufacturer prior to publication.

Since manufacturers frequently change ingredients or formula and/or introduce new and improved products, or fail to distribute in certain areas, trade names are mentioned to help the reader zero in on products of comparable quality and end use. The Publisher

BE AN INSTANT HERO!

A faulty plumbing fixture, a leaky faucet, and a deep seated headache have much in common – your peace of mind.

This book, like aspirin, can relieve many headaches. But unlike aspirin that offers relief, this book does much, much more – it provides cures, repairs that save money.

Those who follow the simplified directions and fix a leaky faucet, or "cure" a running toilet, soon discover the effort works wonders – on inanimate fixtures, as well as on individuals.

Not having to pay outrageous costs for simple repairs, learning how to do something they didn't believe themselves capable of accomplishing, the man of the house takes on glamour, becomes an Instant Hero.

Solving home repairs is an important part of family life. The man who takes care of his home, takes care of his family. And the home reciprocates. While "things" can't say "thanks," they do exert great influence on the lives of those within their sphere of activity.

Don R. Brann

TABLE OF CONTENTS

PLUMBING FACTS OF LIFE

Codes in many cities specify that only a licensed plumber can make certain repairs and fixture installations. Homeowners who want to learn how a repair is accomplished, or insure getting the work done properly, or wish to estimate the time it normally takes, or do it themselves, will find many of the answers by reading this book.

Read slowly. Note each illustration. If you are unfamiliar with plumbing, the illustrations and procedure will naturally seem strange. Don't let this bother you. On some mornings even your face can look strange.

Plumbing isn't nearly as difficult as most people imagine. All it requires is taking something apart—reversing the original process of assembly. If you know where part 1 is located, when, and how to remove part 2, 3, etc., then reassemble parts exactly as each illustration shows, you've got it made.

Every plumbing fixture in your home contains parts that were assembled in position shown in each illustration. To repair, or replace, reverse the process of assembly. Play safe. Number each part, or place small parts on a large calendar, or other numbered sheet. Place the first screw, nut or bolt you remove on 1, Illus. 1.

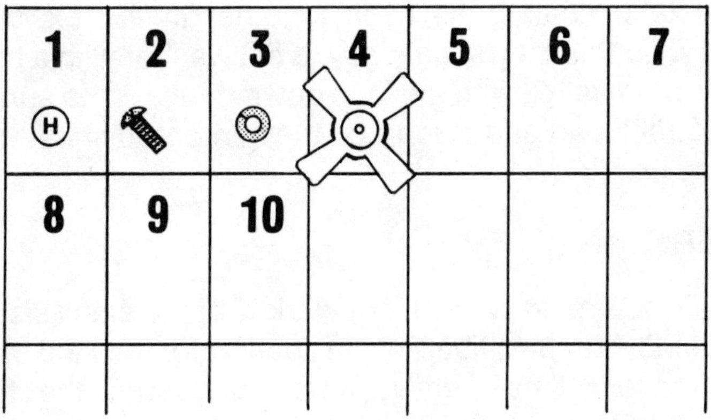

1	2	3	4	5	6	7
8	9	10				

(1)

Through the years, plumbing fixtures have gone through many stages of design. Parts made for one faucet or fixture, seldom fit another. The key to its repair is simple. Just replace each part with its exact duplicate. Learning what part is required, is more than half the task of repair.

A word of advice. Always ascertain the brand name and model number of equipment in your home. Keep this information on file. Continually add to it when you purchase new equipment. When you identify the sink, lavatory, toilet, etc., purchase a set of replacement washers or gaskets. While you may not need these until they are too dried out to use, having replacement samples of what you need, when it's urgently needed, makes life worth living. Being a genius just requires a little foresight.

Always read a service manual prior to purchase of equipment. Manufacturers that have your best interests at heart provide detailed, step-by-step maintenance instructions that simplify making adjustments and repairs. Be sure to use the right name. If it says American, is it just American, or American Brass, American Kitchen, American Standard, etc., etc.

Keep tools required in a convenient place. Set up a vise where it can be used when needed. Remember, one plumbing repair bill can pay for a lot of tools. Two or more repairs, and you make sizeable savings.

Plumbing, like most crafts and trades, depends on sequence. What you take apart goes back together in the exact same position. While this may seem easy, a flat washer with a beveled edge may be installed with the face down or up. To be sure, note position of each part and place it on the numbered sheet with the proper face up.

WASHERS

Washers that "almost fit," seldom work. Get the exact size. Also the exact kind—rubber, fiber, nylon, leather, or the kind that was removed. Invest time getting acquainted with the fixtures described in this book.

2

WASHER CHART

Washer	O.D.
OO	15/32" or 1/2" O.D.
"O"	1/2" or 17/32" O.D.
1/4" S	17/32" O.D.
1/4"	9/16" O.D.
1/4" L	19/32" O.D.
3/8"	5/8" O.D.
3/8" M	21/32" O.D.
3/8" L	11/16" O.D.
1/2"	3/4" O.D.
1/2" L	25/32" O.D.
5/8"	13/16" O.D.
5/8" L (Beveled)	27/32" O.D.
3/4" (Beveled)	7/8" O.D.
3/4" (Flat)	15/16" or 1" O.D.
1" (Flat)	1 9/64" O.D.
1" (Beveled)	1 1/8" O.D.

Faucet washer sizes have been standardized. Illus. 2 shows size and number. This simplifies ordering exact size. Buy hot and cold washers that fit. Buy a few replacement screws.

The washers shown on pages 136 to 152 are as close to full size as reproduction permits. Knowing what to ask for, helps your retailer find what's needed.

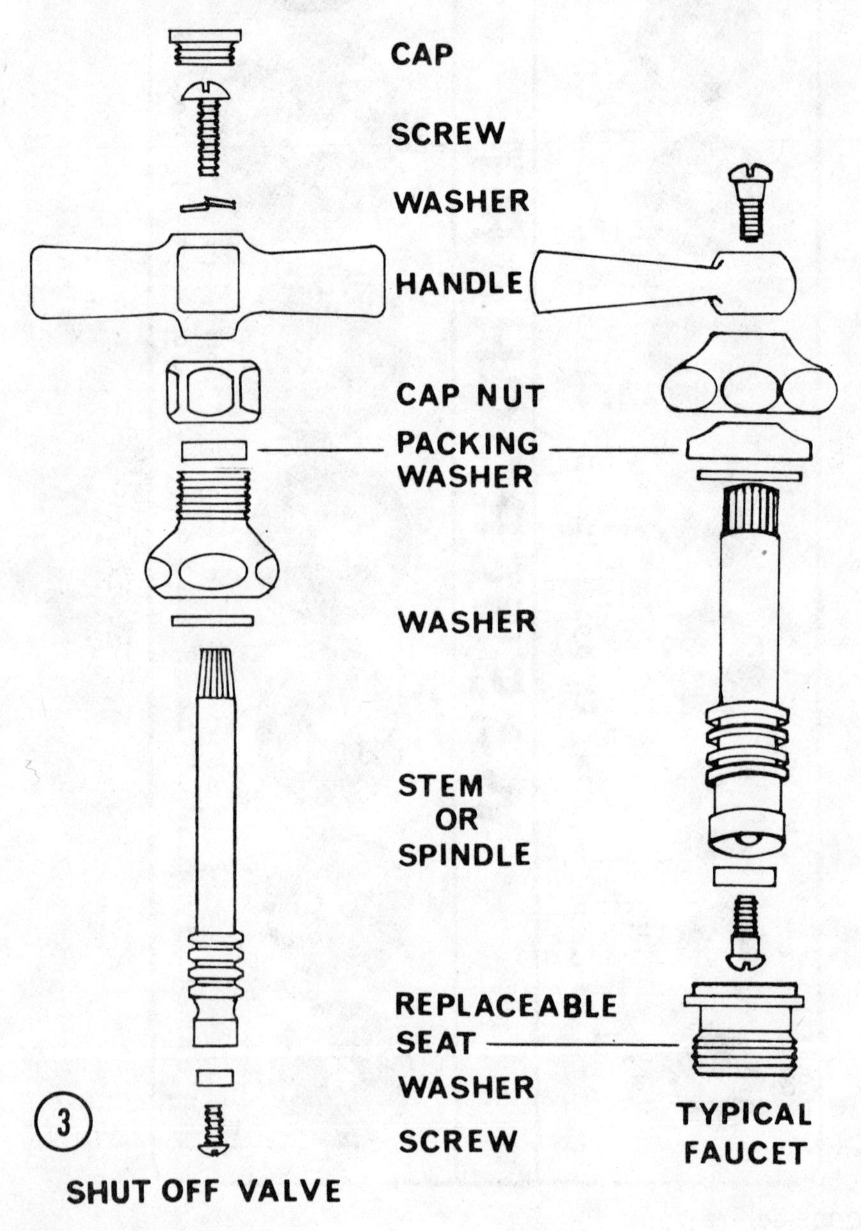

To play safe, remove the screw holding washer, Illus. 3, to see if it is still removeable. If not, a few drops of penetrating oil, (Liquid Wrench or equal) should loosen it. Or purchase a replacement stem, Illus. 4.

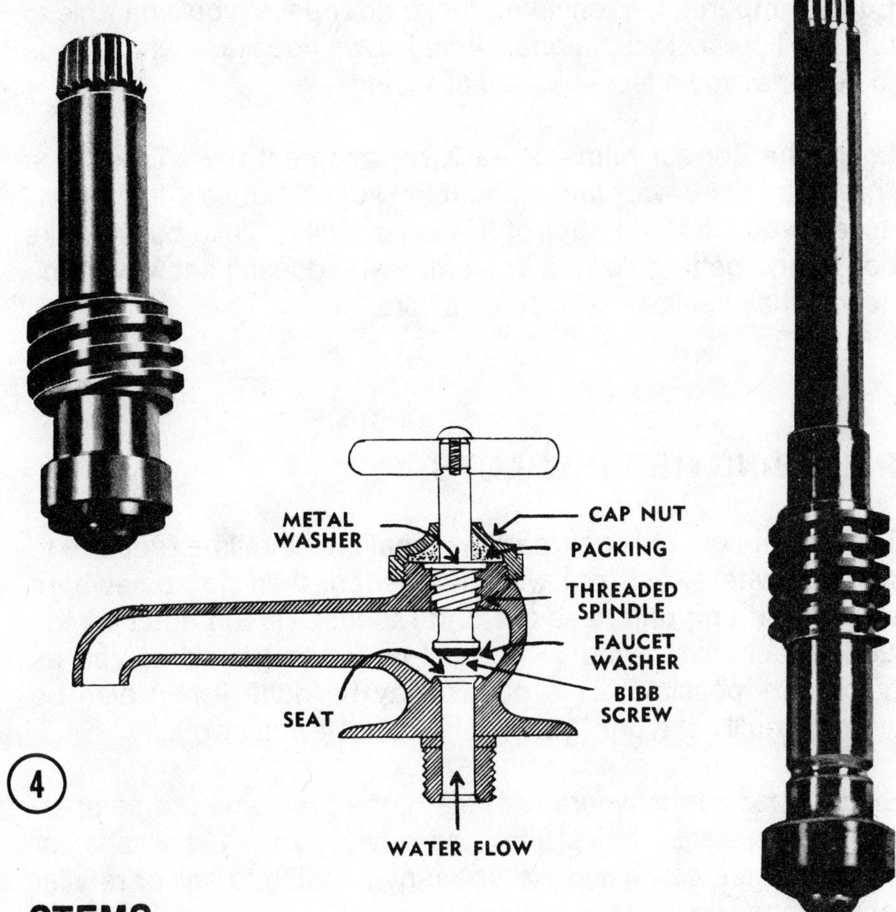

STEMS

Through the years manufacturers of sinks and lavatories installed dozens of different types of faucets that are no longer in production. The stems shown as close to full size as reproduction permits on pages 104 to 127, help identify a replacement. After removing a stem, match it up, then phone your plumbing supply retailer to find out whether they have a replacement. Always take along the old stem. It is advisable to replace old seat. The seat number (in most cases) is shown alongside the stem. Pages 128, 131 show seats full size.

Manufacturers assemble products according to their specialized methods, not according to industry standards. For this reason, step-by-step directions describe and illustrate basic repairs. Use it as a guide, and you'll soon know a ballcock from a toilet seal. Equally important, even if you make no repairs, you'll be able to talk intelligently to a plumber. If he knows you know what needs to be done, you effect substantial savings.

Learn the correct name of each replacement part. Talk like a "pro" when you walk into a plumbing supply house and they will accept you as a member of the inner circle. While busy clerks don't mind getting rid of a customer who doesn't know what he needs, they hesitate to louse up a "pro."

PLUMBING REPAIR FACTS

The actual cost of repair parts is small. It's the time required to go somewhere, find out what is needed, then go somewhere else to find and purchase the part required. If you don't tell the clerk exactly what you want, he'll guess to get rid of you as quickly as possible. This guess may be right, it can also be wrong. You'll get a part, you'll put it in, but it won't work.

Since other homeowners with the same problems are competing for the sales clerk's time and advice, your 10¢ washer or $2.00 gadget, can only take so many minutes to sell or retailer loses money on the transaction.

If you live any distance from a well stocked plumbing supply house, or it has more customers than its staff can conveniently serve, write the manufacturer. Tell him what equipment you own. Order a kit of repair parts. Do this before you need it.

Never attempt a plumbing job when you are tired, angry, or lack the time. If you "fear" doing it—that's OK. Always fear doing something you have never done before and you exert that extra something that keeps you alert.

The plumbing in your home represents an important part of your total investment. Builders allocate 10% or more of the overall construction costs for plumbing. It can cost you double to replace plumbing after the house is completed.

As head of a household, it is your responsibility to establish operational rules for all equipment, and see to it they are followed. A waste paper basket should be first on your list for bathroom maintenance. Or if space permits, a small covered waste receptacle. Insist that all paper towels, bandaids, bandages, sanitary napkins, cigarette wrappers, cigar butts, etc., be deposited in the wastepaper basket, not down the toilet.

Many adolescents vent jealousy and temper by deliberately throwing something down a toilet. Plumbers report the highest number of house calls in homes where teenagers entertain.

Learn where waste and water lines are located. Where to find cleanout plugs. This information can save you both time and money.

Don't allow anyone to drive nails in a wall that contains pipes. Copper tubing is especially easy to puncture. Since water will invariably follow a framing member, the actual leak frequently shows up some distance from its source.

If you find a leak where there's no fixture or pipe nearby, first ascertain whether anyone remembers driving any nails into a wall, and where it was done. In many cases, nails driven weeks or months previously can provide an important clue, even when the actual leak occurs far away.

The wall framing, Illus. 5, behind most kitchen and plumbing fixtures, contains 2x6 studs, placed 16" on centers. Studs may be notched to receive tubing, or holes drilled as shown, Illus. 5. Cats are usually nailed in position shown to support frame.

A nail frequently acts as a plug when driven into copper tubing. A leak can start months later, when the nail begins to rust. When you remove the nail, it accentuates the leak. If you find a nail driven into a wall containing copper tubing, use a screw driver to scrape out a small hole in plaster. If your sleuthing is accurate, you'll get water in your face.

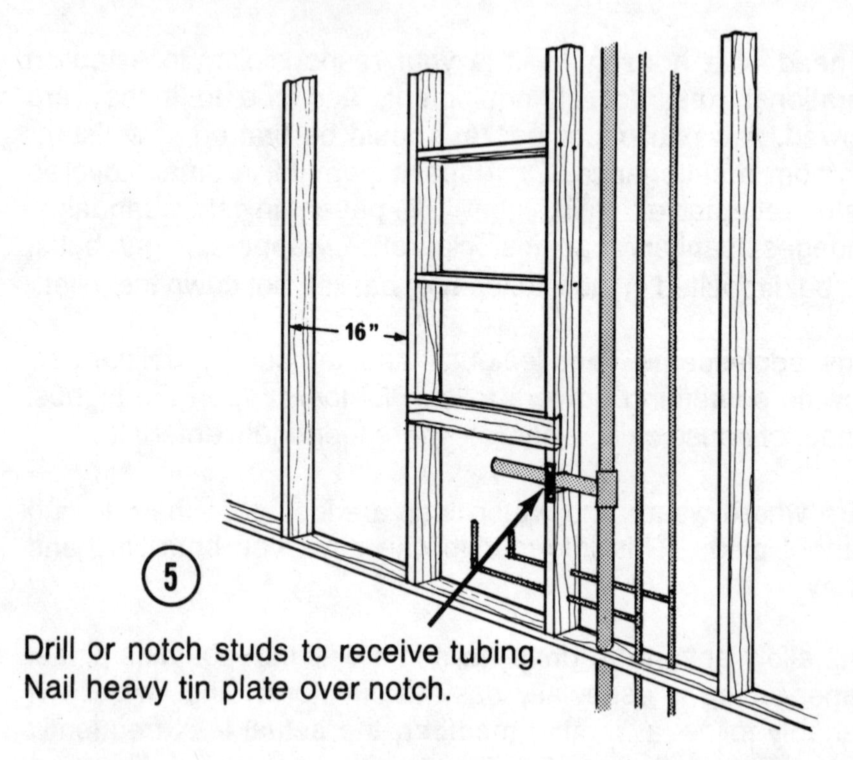

Drill or notch studs to receive tubing.
Nail heavy tin plate over notch.

Fix hole in pipe as specified on page 77. Patch hole with patching plaster. Paint wall. Always hang pictures using paste-on hangers.

SEPTIC TANKS

Those who buy homes in a sub-division, or in any area served by septic tanks should obtain a drawing, showing A, length and location of line between house and tank; B location of tank; C direction and length of drain fields, Illus. 6.

Also ascertain position of intake pipe in septic tank, Illus. 7. This information can save considerable money and labor when it's necessary to service tank.

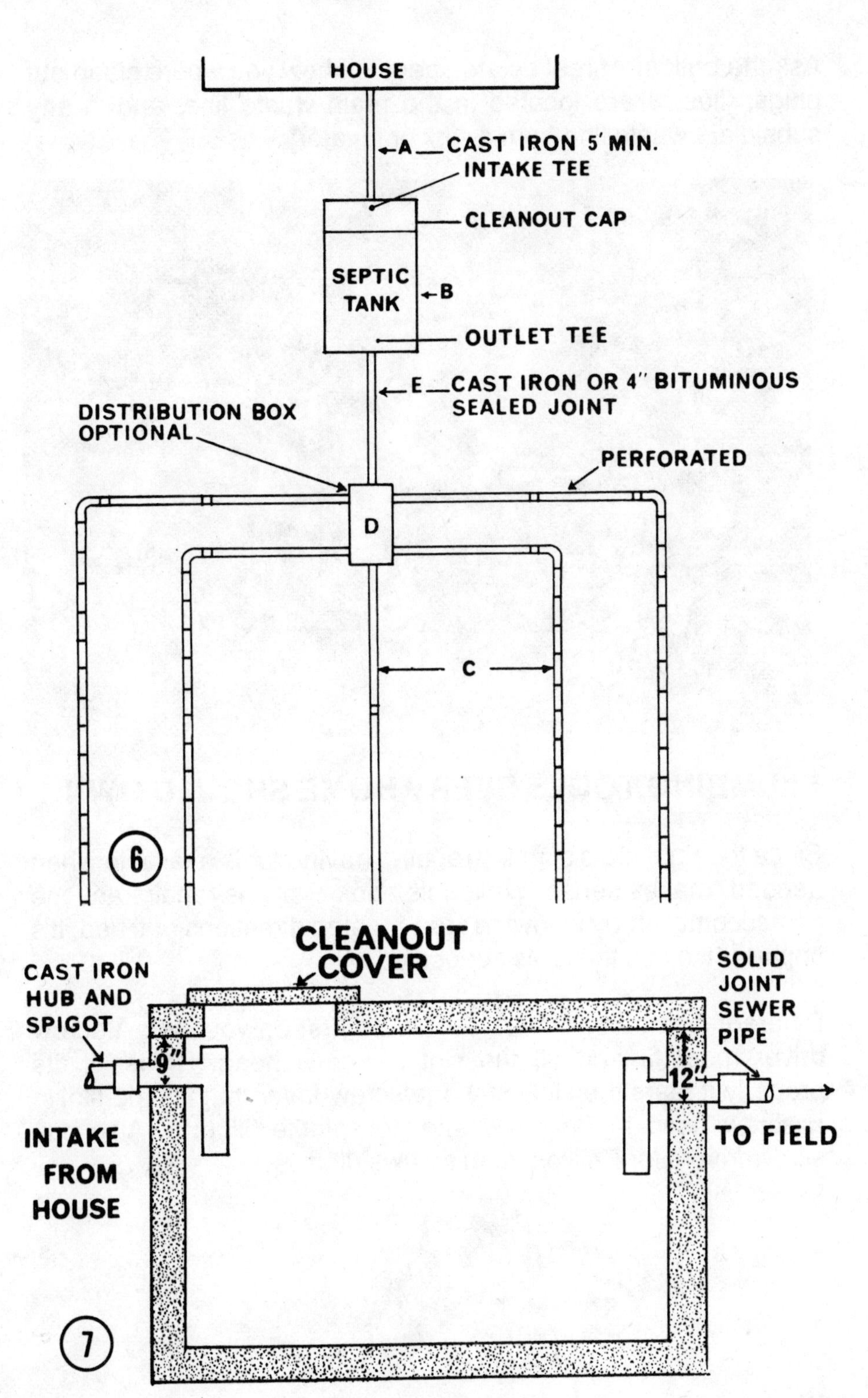
HOUSE
A CAST IRON 5' MIN.
INTAKE TEE
CLEANOUT CAP
SEPTIC TANK
B
OUTLET TEE
E CAST IRON OR 4" BITUMINOUS SEALED JOINT
DISTRIBUTION BOX OPTIONAL
PERFORATED
D
C
6
CLEANOUT COVER
CAST IRON HUB AND SPIGOT
9"
SOLID JOINT SEWER PIPE
12"
INTAKE FROM HOUSE
TO FIELD
7

Ask the builder or real estate agent to show you where clean out plugs, Illus. 8 are located in the main waste line, and in any subsidiary waste line from a sink or lavatory.

PLUMBING TOOLS EVERY HOME SHOULD OWN.

Since the right tools simplify repairs, having tools available when needed, makes sense. While this book explains repairs anyone can accomplish by following step-by-step directions outlined, it's important to use the tools suggested.

Different size screwdrivers should be first on your list. A screw driver must fit and fill the slot in screw head. Unless it fits properly it tears the slot. Buy one screwdriver that fits the slot in a screw used to fasten washer to spindle, Illus. 3. Also buy screwdrivers for Phillips head screws, Illus. 9.

Buy a can of penetrating oil, Liquid Wrench or equal. This now comes in spray, as well as in spout cans. The spray can is handy in hard to reach areas. Penetrating oil saves fixtures, parts, time and temper. Don't force anything. Use a little oil. Give it time to penetrate then tap the part gently to break a rust bond. Next use the tool required.

A stillson wrench Illus. 10, is handy in many places. It's especially handy if you don't happen to have a small vise. While a stillson or a pipe wrench isn't used often, it is one tool needed to hold pipe.

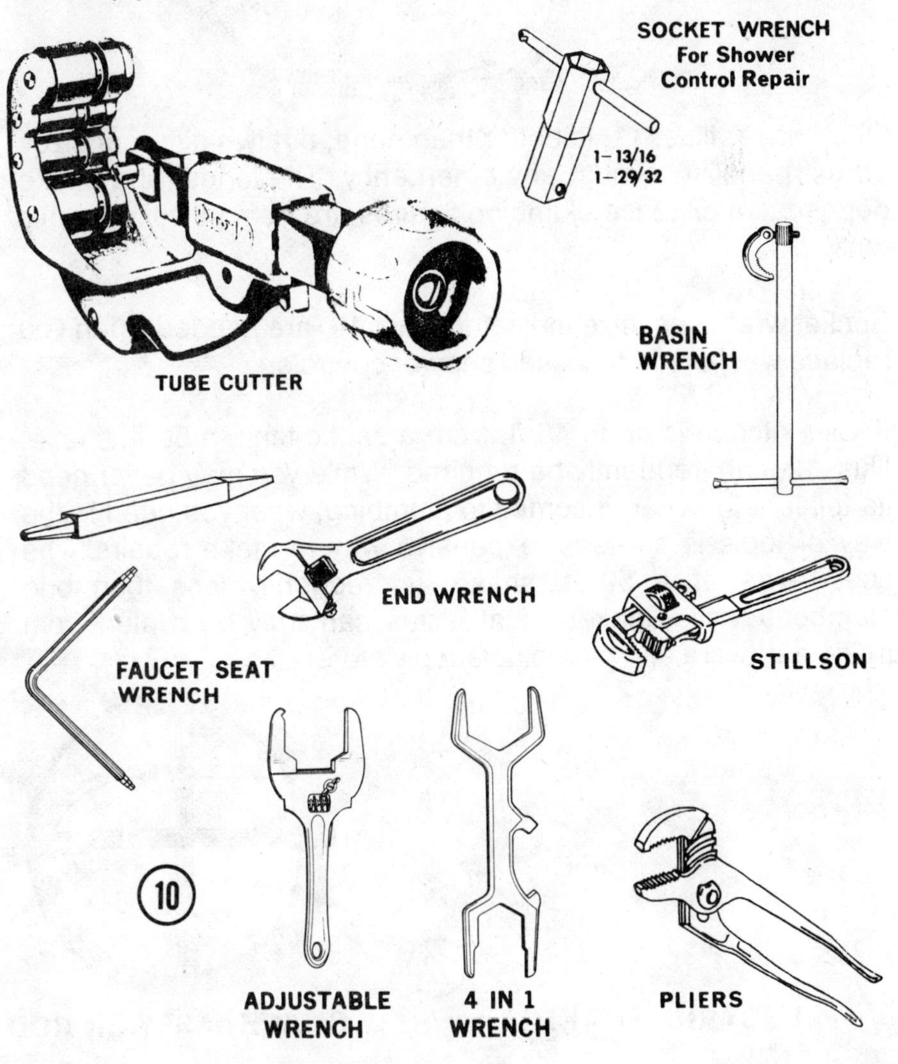

Long handle pliers can sometimes be used in place of a stillson, if you are able to apply the strength required.

Two different size adjustable end wrenches are recommended. A roll of adhesive, or electric tape should also be kept handy. Wrap tape around a polished fitting before applying a wrench and you save the finish.

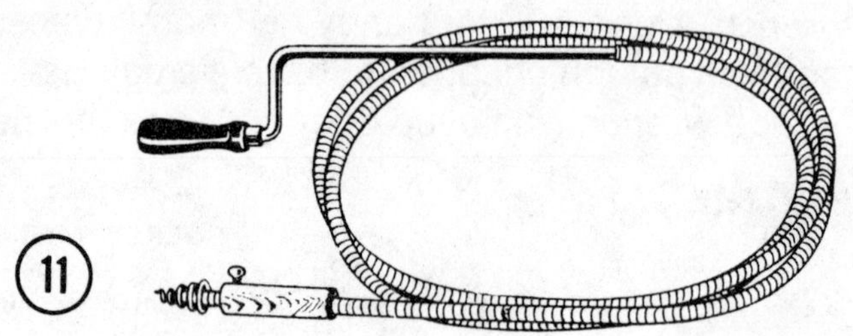

(11)

One snake, Illus. 11, is better than none, but two different sizes permit handling almost any emergency. The length of a snake depends on distance plumbing fixtures are from sewer or septic tank.

Socket wrenches, size indicated Illus. 10, are needed when you replace washers in recessed shower controls.

If your kitchen is up to 45 ft. from a septic tank, a 50 ft. snake, Illus. 12, can frequently be required. While you may never need its full length, when it comes to plumbing, what you need in the way of tools is the least expensive way to make repairs. The cost, even of a 50 ft. snake, is frequently less than one plumber's visit. Snakes of all sizes can now be rented from many hardware and plumbing supply stores.

ELECTRIC SNAKE

(12)

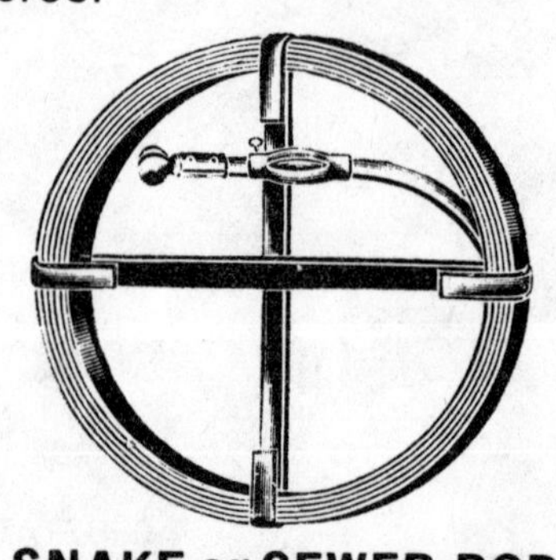

SNAKE or SEWER ROD

A short snake, also called a closet augur, Illus. 13, is ideal for opening clogged toilets. The toilet bowl is shaped as shown with some variation depending on manufacturer.

Before inserting augur in toilet,pull handle all the way up. Insert head in bowl, push rod down while you turn handle.

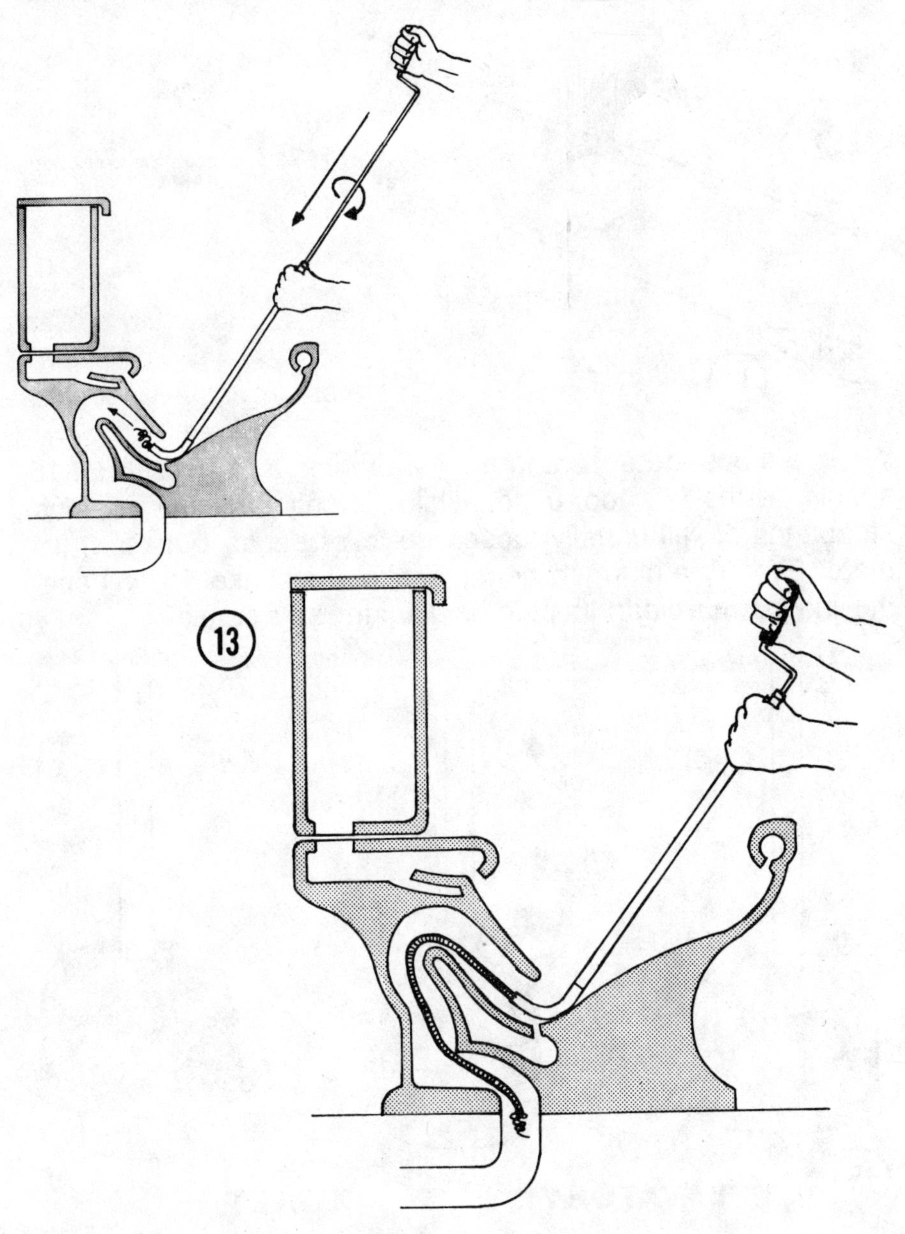

Using pliers, cut a wire coat hanger and straighten it out. Make an eye loop at one end, Illus. 14. You can frequently loosen and retrieve a comb, toothbrush, ball of paper, or other matter that never should have been dropped into a toilet. Keep turning the wire. Use care not to scratch finish.

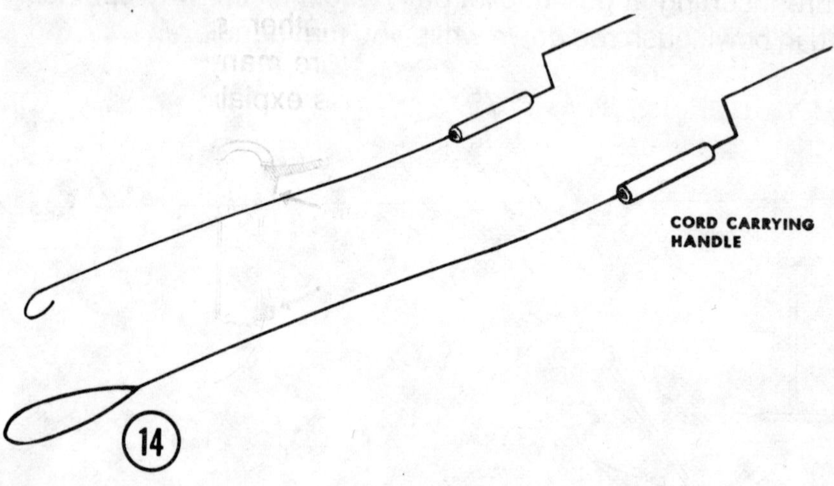

When a stoppage occurs in a toilet or sink, a plunger, Illus. 15, should be the first tool used. While it's only effective for minor stoppages, it will usually loosen up toilet paper, but not much more. First try a plunger, next a snake. If a snake doesn't open the line, open a cleanout plug, and again use a snake.

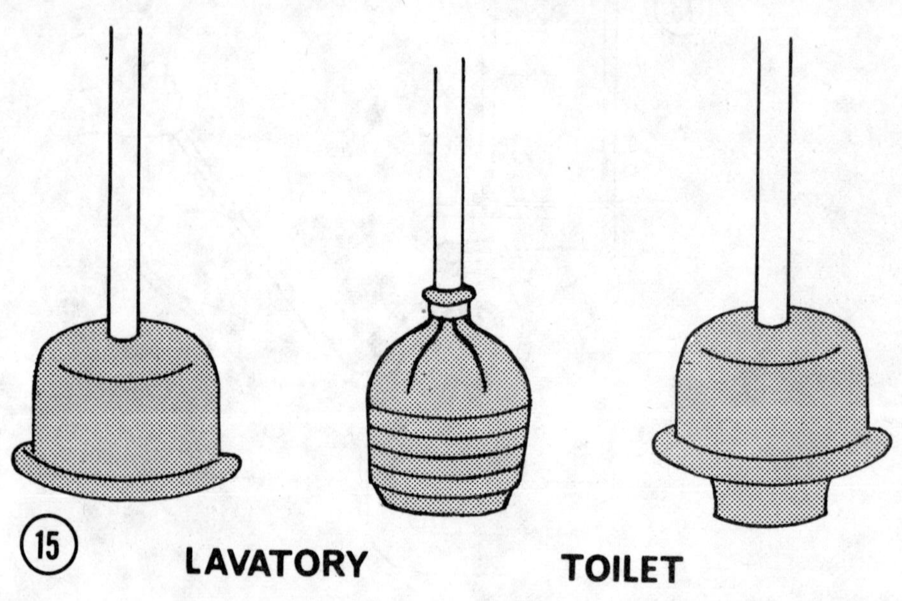

FAUCET REPAIRS

The first step in repairing a leaky faucet, sink or basin, is to shut off water either below fixture, Illus. 16, or at water meter, Illus. 17. Open and drain faucet. Place stopper in drain, or cover drain with a rag to prevent a screw or other small part from disappearing down drain. Since there are many different types of faucets and shutoff valves, directions explain two types that contain most component parts.

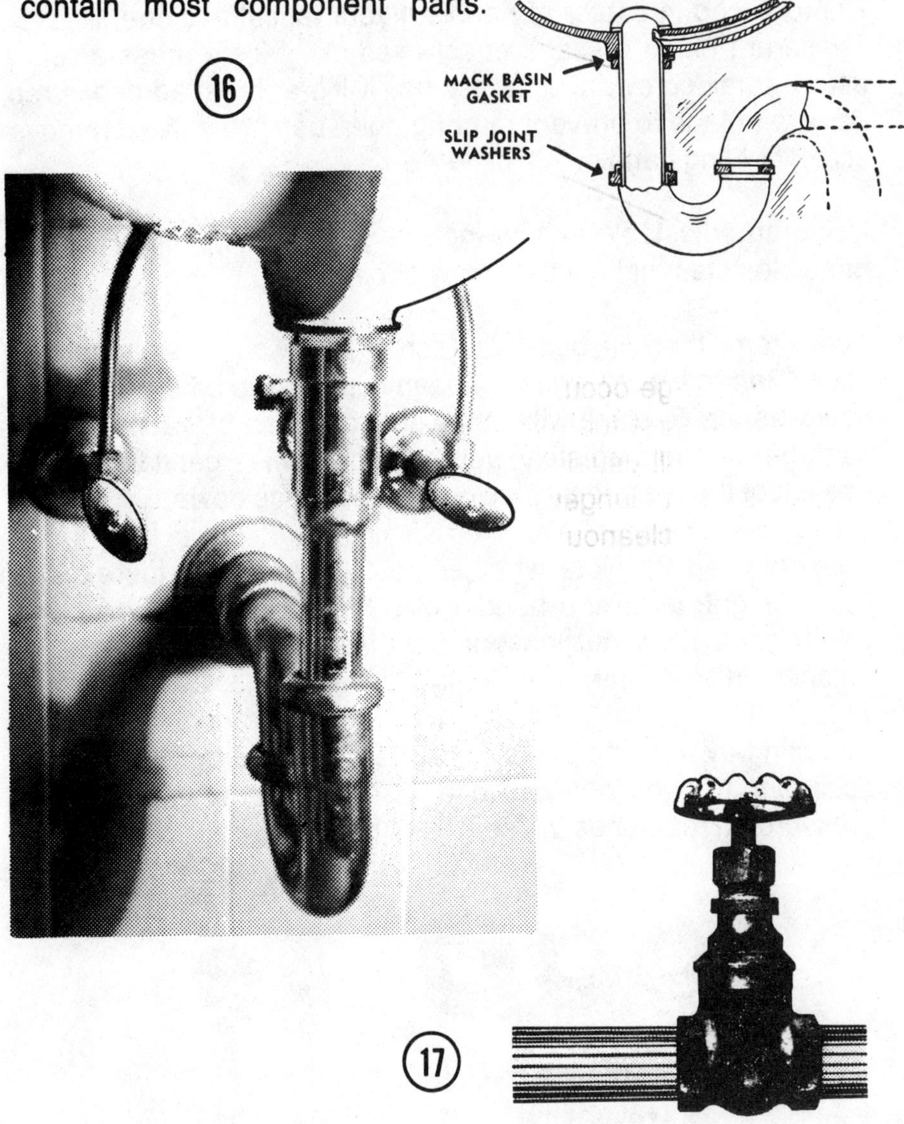

Illus. 18 shows sizes of replacement mack basin, sink and bath gaskets.

(18)

Part No.	Description
G-101	(1¼″) BASIN (1¼″ I.D.)
G-105	(1½″) BASIN (1 7/16″ I.D.)
G-108	(1¾″) S & B (1 13/16″ I.D.)
G-109	(1⅜″) S & B (1⅜″ I.D.)
G-111	(1¼″) WIDE P. (1¼″ I.D.)

Remove cap in center of handle if your faucet has one, Illus. 3. These are either pressed or screwed on. If it's a press-on cap, use a small screwdriver to pry up. If it's a threaded cap, wrap edge with tape to prevent marring, then use pliers. Next remove screw holding handle.

Using an adjustable end wrench, remove packing nut. Replace handle on stem but not the screw or cap.

Turn stem, then lift out. Note screw and washer at bottom of stem. Remove worn washer. Clean cup or base before installing new washer. Replace with same size and kind of washer. If you can't get one immediately, you can sometimes get a little more use out of the old one by placing the good face down.

Look into faucet. Note whether seat is smooth. If there's any scale or grit, wrap the handle of a toothbrush with a rag and polish seat. Blow out loosened scale. If you see any cuts or notches on seat, ream seat lightly.

You can buy a reamer, Illus. 19, at hardware stores. Use reamer according to directions provided. Use care not to put too much pressure on reamer as the seat is soft.

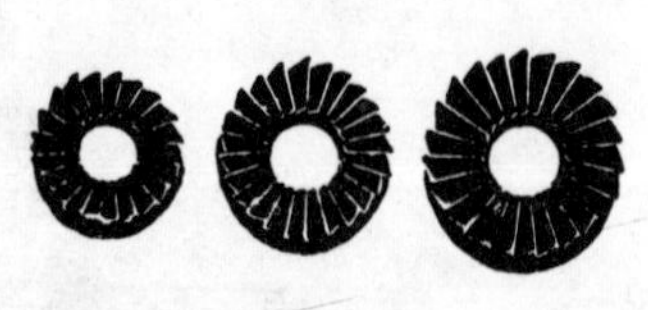

(19) Use proper size cutter in seat reamer.

If a seat can't be reamed smooth, replace with a new seat, Illus. 20. This requires the wrench, Illus. 21. Or you can use a slip-on seat. These are pressed over present seat, Illus. 22. If you don't replace a seat, clean rust or grit off existing seat before pressing on a slip-on. See pages 128 – 131 for detailed information on seats.

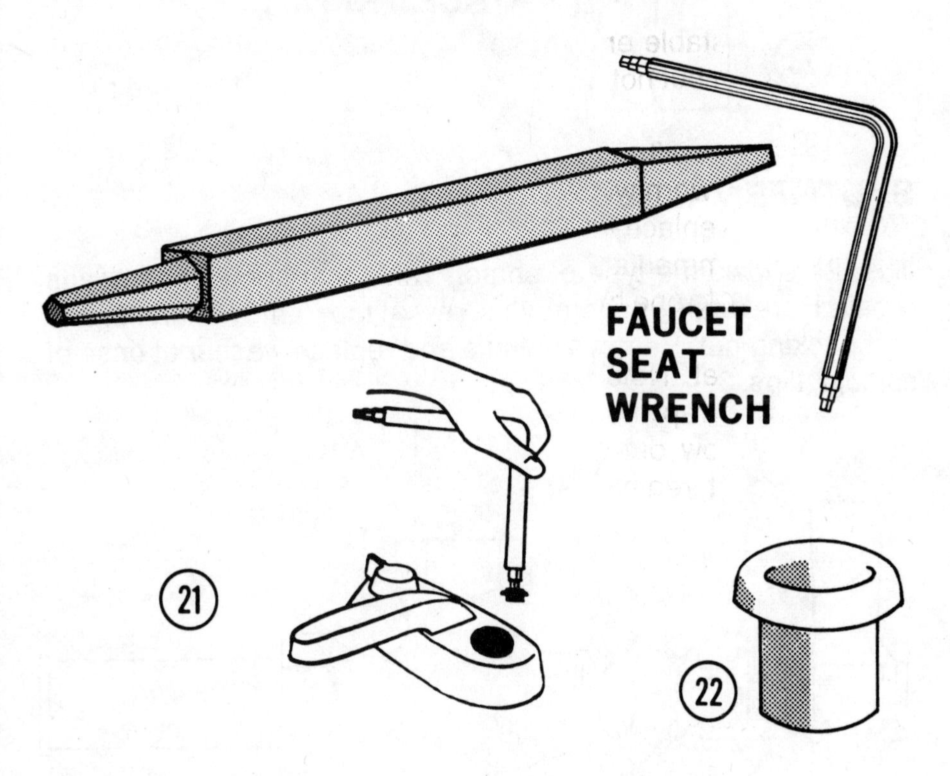

After replacing washer, fasten screw. Replace washer under packing nut if stem had one when you took it apart, Illus. 3. Apply vaseline to threads on stem, replace stem. Tighten packing nut.

If you can't get a packing nut washer, wrap stranded graphite asbestos wicking around spindle, Illus. 23. Turn packing nut down snug against wicking and it compacts into a watertight washer.

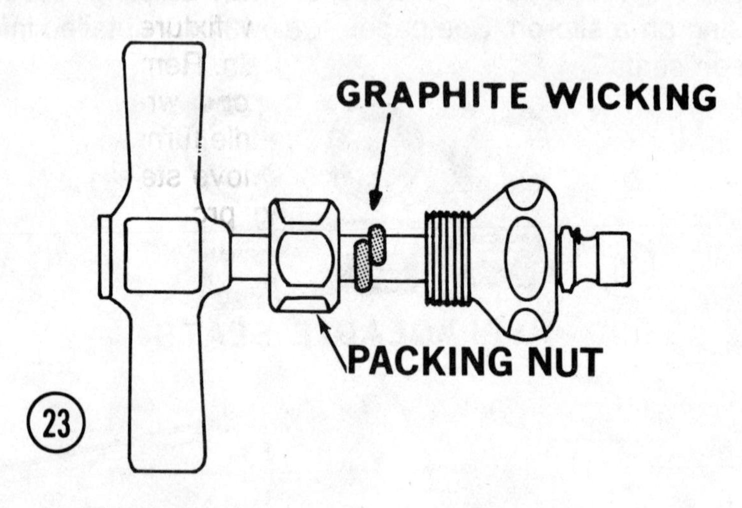

SHUTOFF VALVE

Illus. 24 shows the type of shutoff valve that sometimes needs repair. Here again you remove screw at top, remove handle, cap and packing nut. Remove spindle and replace washer at base of spindle, Illus. 3.

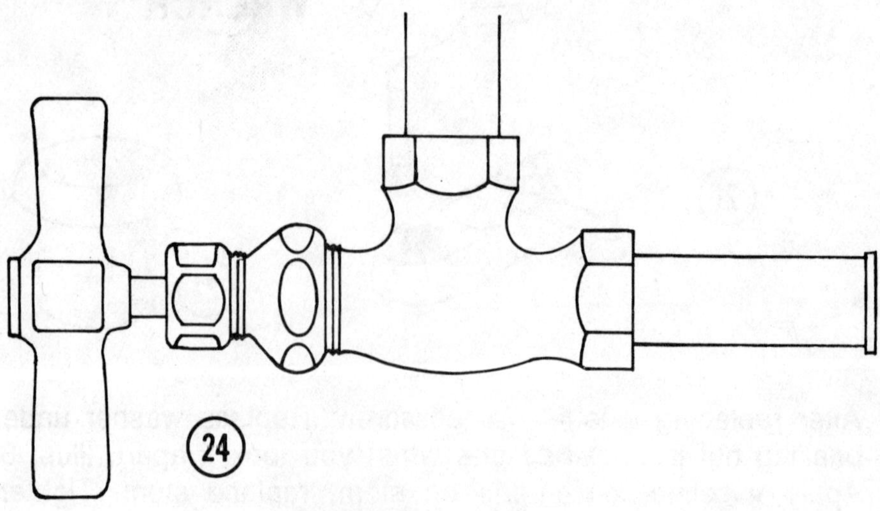

SHUTOFF VALVE

FAUCET REPAIRS AND REPLACEMENTS

Handles on deck faucets, Illus. 25, 26, 9 , on sinks and lavatories, are frequently fastened with a Phillips head screw, Illus. 26. Shut water off at valve below fixture. Open faucet halfway. After water drains out, close drain. Remove screw and gently pry up handle. Using a large end wrench, unscrew packing nut in same direction faucet handle turns. Keep turning stem while you loosen packing nut. Remove stem, packing nut and sleeve. Replace washer following procedure previously outlined.

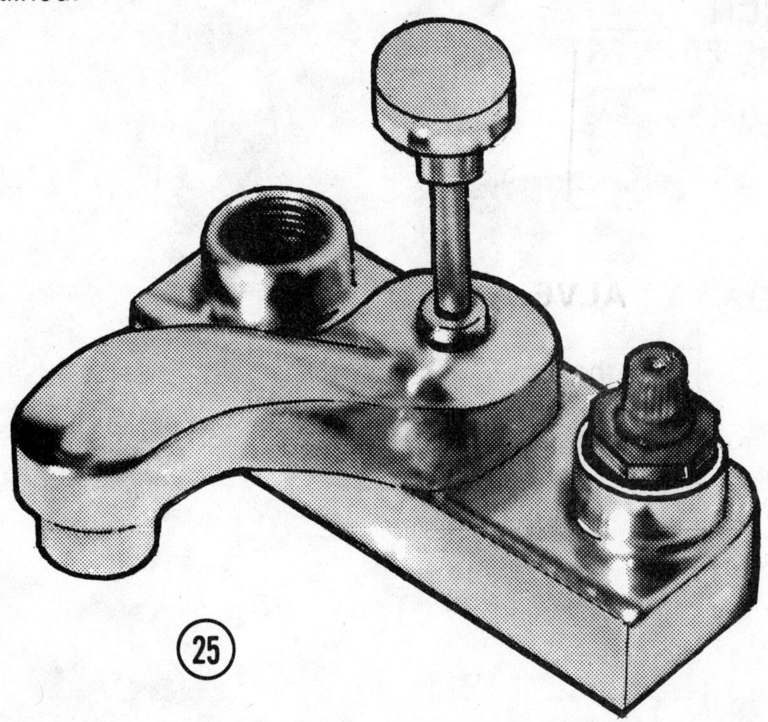

If you need to replace a deck faucet, measure spacing between stems. These vary from 4", 6" to 8".

If water comes out of faucet where stem enters packing nut, tighten packing nut. If this doesn't stop leak, remove handle and packing nut, replace washer under nut or use graphite wicking, page 24. If you can't make a satisfactory repair, consider replacing the outmoded faucet with a replacement. These are easy to install.

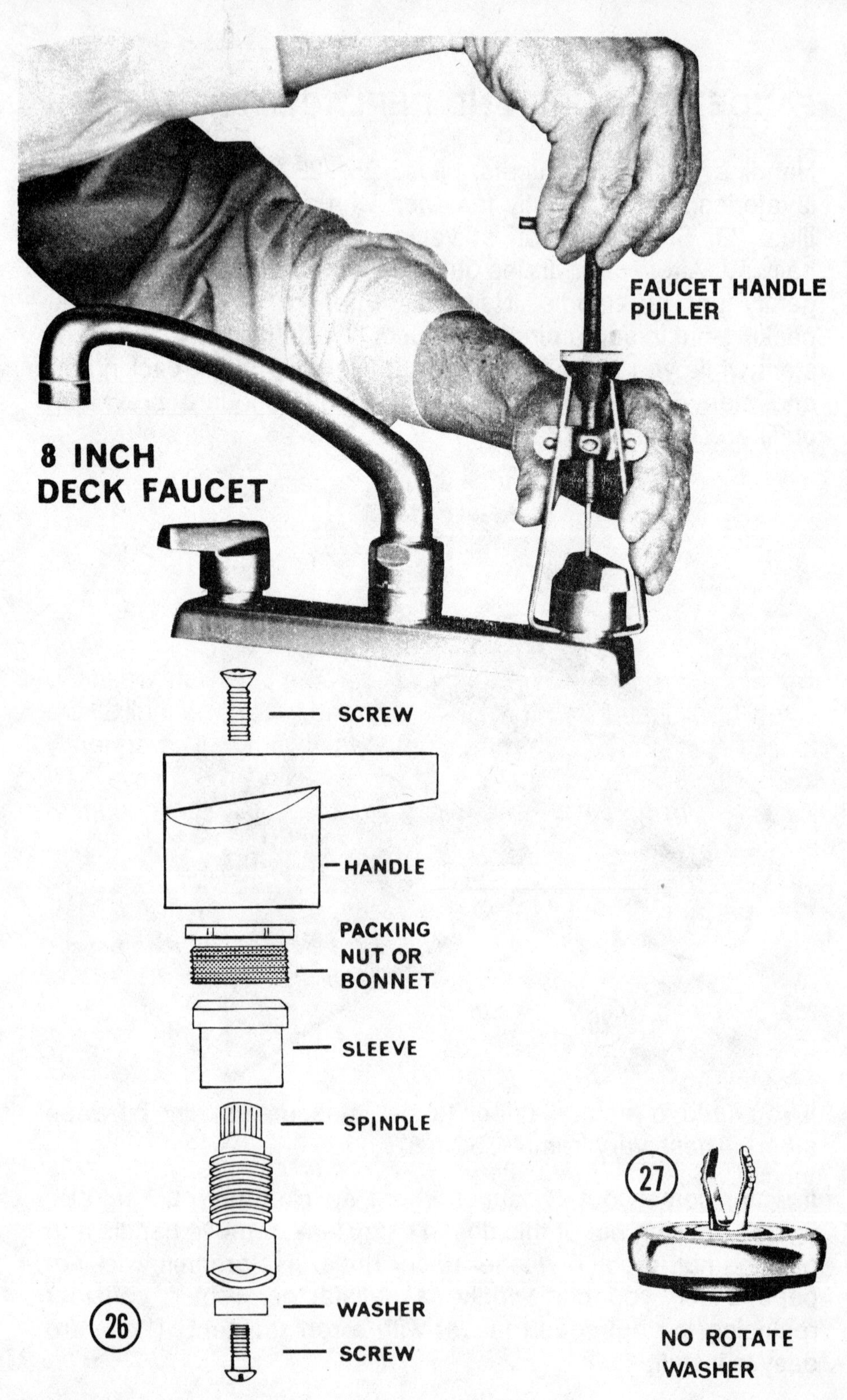
FAUCET HANDLE PULLER
8 INCH DECK FAUCET
SCREW
HANDLE
PACKING NUT OR BONNET
SLEEVE
SPINDLE
WASHER
SCREW
26
27
NO ROTATE WASHER

When a faucet fails to provide water when turned on, the problem can usually be traced to a faulty washer in the faucet, or one in the shutoff valve serving the line. Shut off water. Remove faucet stem, Illus. 26. If washer is loose or remains in faucet, you have discovered the cause. If washer remains in faucet, use thin, long-nosed pliers to pull washer out. Replace washer and screw.

If same can't be obtained, use a No-Rotate, or equal swivel head washer, Illus. 27.

No-Rotate or swivel head washers, are available in sizes that fit most stems. Select the correct size and press into end of stem.

No-Rotate washers lock into position and provide a simple solution to a troublesome problem.

If you have difficulty removing shank of a screw, spray or soak end of stem in Liquid Wrench. After allowing oil to penetrate, tap screw head lightly and turn. If screw head breaks off, drill a hole to size No-Rotate washer stud requires, Illus. 27. It's frequently easier to replace the stem. Match the stem you remove with those shown on pages 104 to 127. To play safe, take the stem removed with you when you go to store.

When a faucet can't easily be repaired with a washer or a new stem, don't fight it. Replace the faucet. It's no big deal and costs much less than most service calls. Relatively few homeowners realize their home represents one of the soundest savings bank they will ever have the opportunity of investing in. As the population continues to explode, and good houses or building sites get scarce; as wages rise and work rules curtail output per man-hour, your home continually increases in value while it pays dividends in better living. How you maintain your home can pay a big bonus the day you decide to sell.

A case in point is the kitchen sink. It's one of the more important conveniences in your home. If your faucets have had it, replace them with either a handsome, single handle washerless unit,

Illus. 28, with a separate spray, or a two handle replacement, Illus. 29. Besides being a joy to use, each updates your sink. A kitchen sink is one of the first things a prospective buyer inspects.

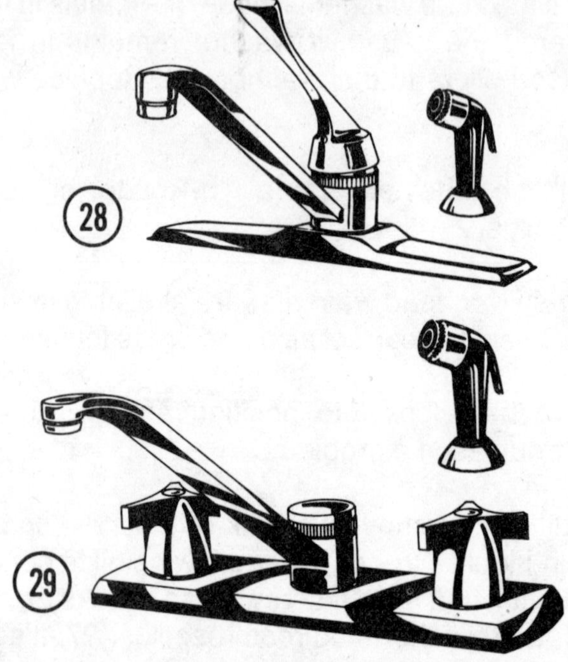

Prior to purchasing a replacement, look under the sink with a flashlight. Measure distance, center to center, Illus. 30, between the studs. Does it measure 6" or 8". Measure distance from stud to supply line. This provides two important measurements required when buying a replacement.

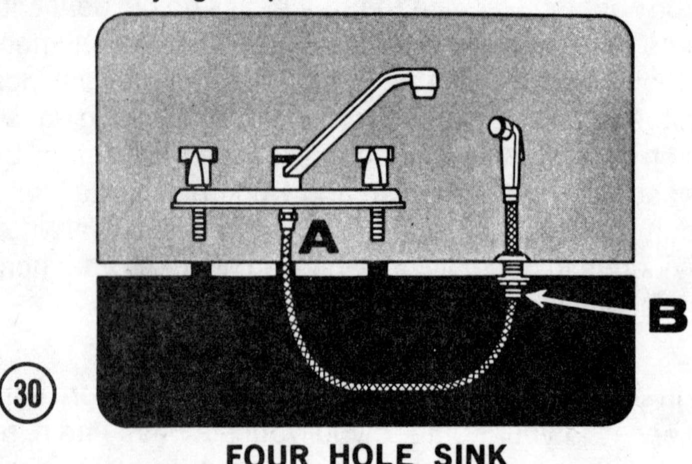

FOUR HOLE SINK

Replacement faucets are available for three and four hole sinks, Illus. 30, 31.

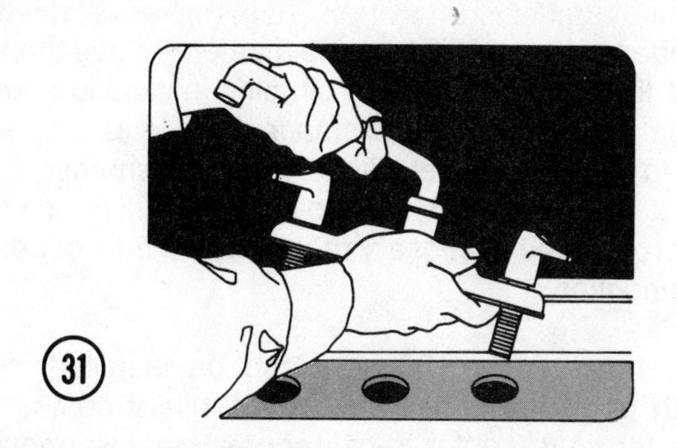

Illus. 32 shows a faucet without a spray attachment. Illus. 33 shows a faucet with spray in base. This fits a three hole sink, Illus. 34.

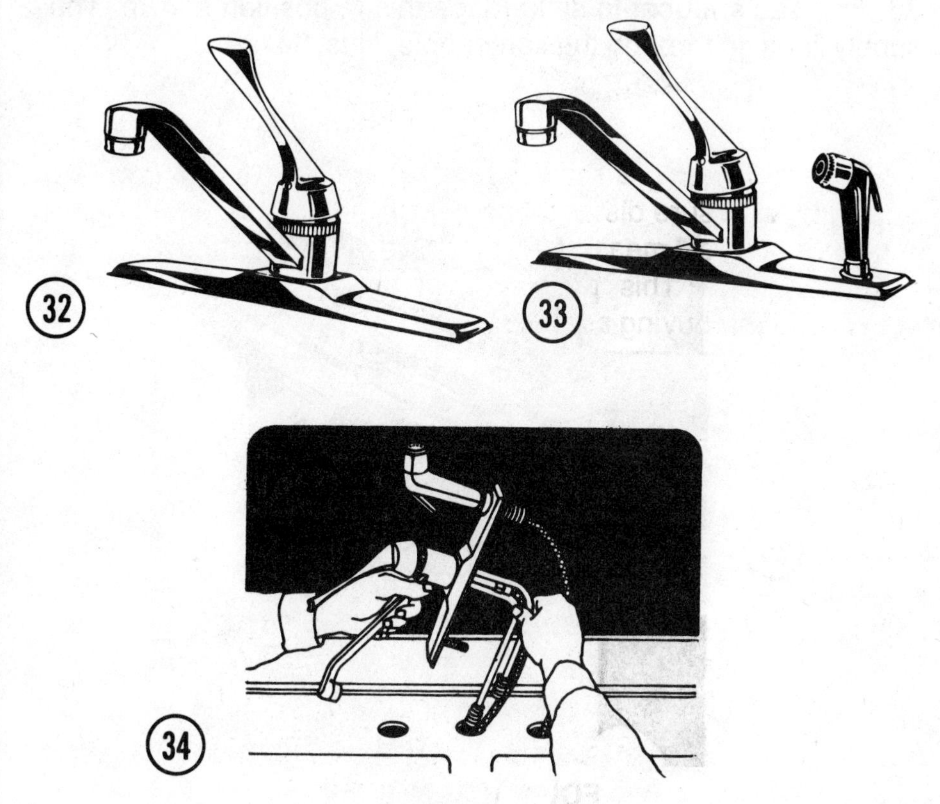

Illus. 28 shows a single handle faucet with a separate spray. This requires a four hole sink.

The 3/8" copper supply tubes (outside diameter) have threaded adapters that fit 1/2" cast iron. To simplify installation, and to buy what you need, measure distance from end of present supply lines to existing faucet. Regardless of whether supply line comes out of the floor or wall, knowing exact distance helps your retailer provide the necessary replacement, and/or adapter to make a connection.

Replacement faucet, Illus. 28, 29, can be installed in a 4 hole sink with openings either 8" or 6". While it comes from factory set up for an 8" center-to-center opening, by removing 3 screws, and moving lugs to inside holes of a cleverly designed mounting plate, the mounting bolts fit holes spaced 6" apart. This replacement unit comes with a black base gasket, Illus. 35, 36, that seals faucet to sink. Place this in position shown. The supply lines go through the center hole, Illus. 34.

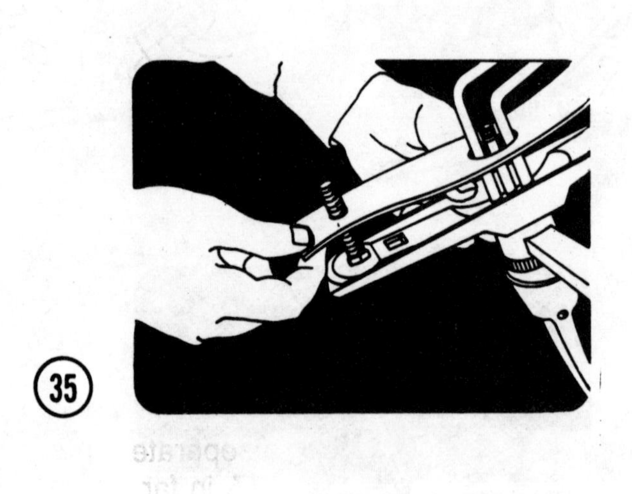

35

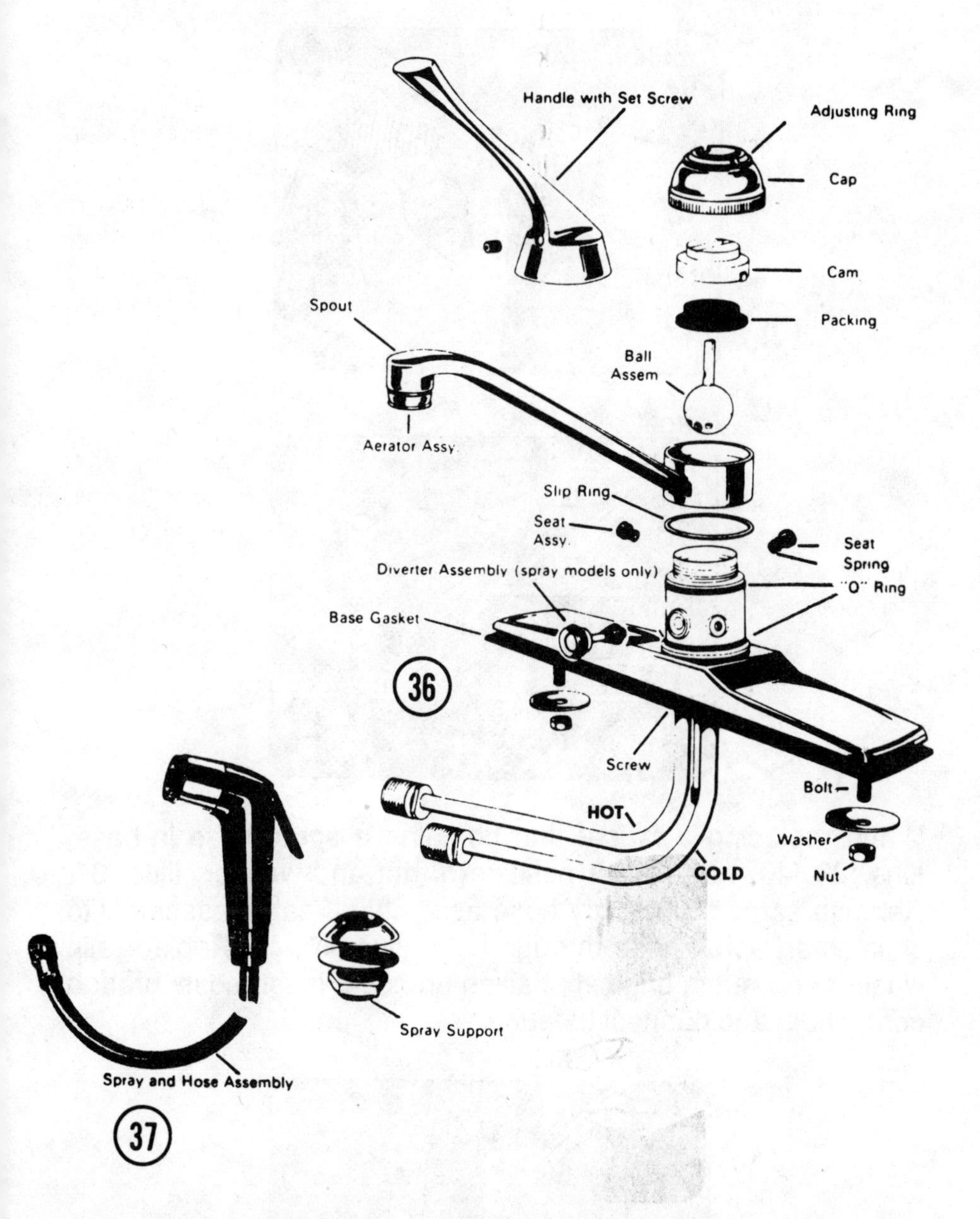

If you are replacing a faucet that has a separate spray, Illus. 30, fasten spray support assembly, Illus. 37, in far right hole, Illus. 38, 39. Insert spray hose down through fitting, then up through center hole. Apply pipe dope or Teflon plumbing tape to threads, (if faucet manufacturer recommends same) and fasten spray hose to nipple.

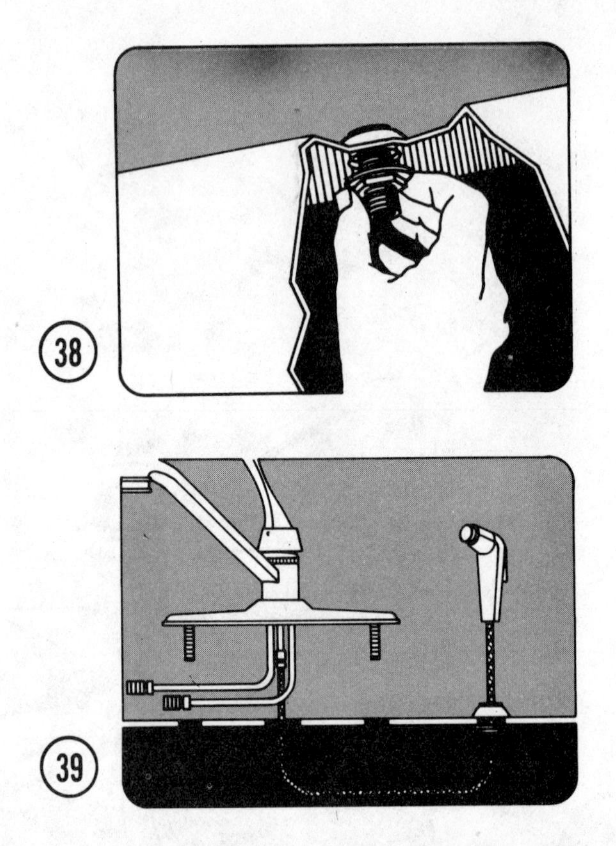

When replacing a faucet that contains a spray hose in base, Illus. 33, 40, note exact position of nut and washer, Illus. 37. Remove same from spray hose assembly. Fasten assembly to sink. Insert spray hose through hole, Illus. 40, 41. Replace slip washer and nut in original position on hose. Insert hose through center hole and connect to faucet.

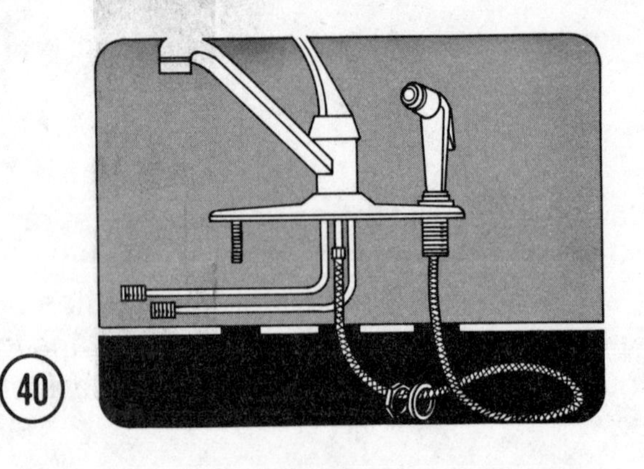

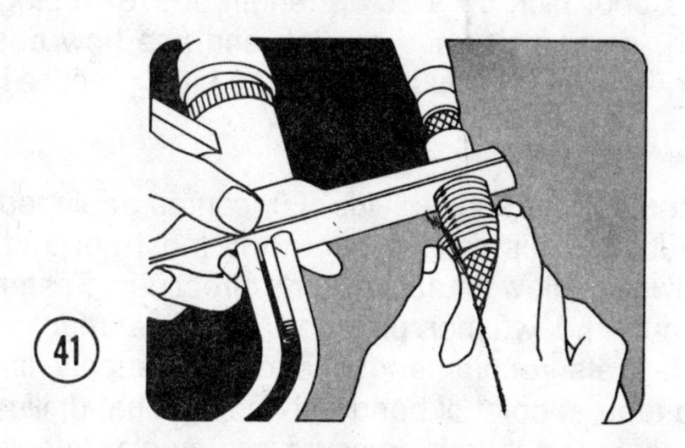

Without bending supply lines, insert supply tubes in center hole, Illus. 34. Position and fasten faucet to sink with mounting nuts and washers. Use extreme care not to kink the tubes or you'll be dead insofar as the faucet guarantee is concerned.

To bend tubes to meet supply lines, always consider yourself facing the mounted faucet. Bend tubes to meet proper supply lines.

The faucet shown comes from factory with right hand copper tube hot, left hand tube cold.* To connect these to most supply lines, bend tubes to shape shown, Illus. 42, 43, 44.

* The hot water supply is usually on the left in most existing installations.

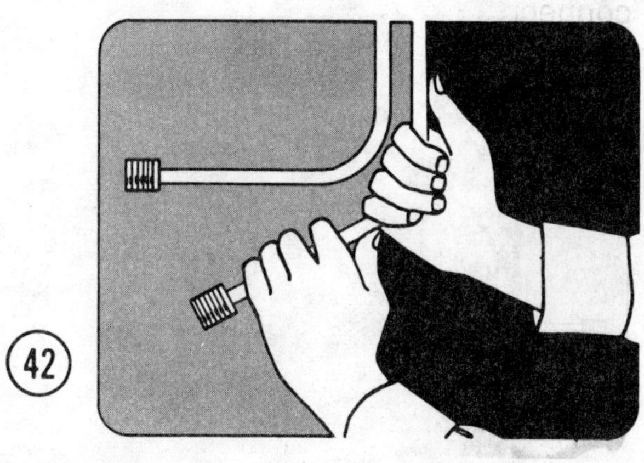

If you have never bent copper tubing, and don't want to make an expensive boo boo, pick up a scrap length of 3/8" tubing in a service station, garage or used car lot, and see how easily it bends, and/or, how easily you can kink it. A bit of practice helps make perfect.

The replacement faucet shown, Illus. 28, comes equipped with 3/8" copper tubing. This can be bent to fit most hot and cold water lines. Always follow manufacturer's directions. Fasten unit to sink with nut and washer provided. The manufacturer of replacement faucets recommends bending a supply line by firmly gripping tube at point of bend with the right hand, Illus. 42, and applying palm and thumb pressure as you slowly pull tube into position with your left hand, Illus. 43, 44. This, of course, is done after faucet assembly has been secured to sink.

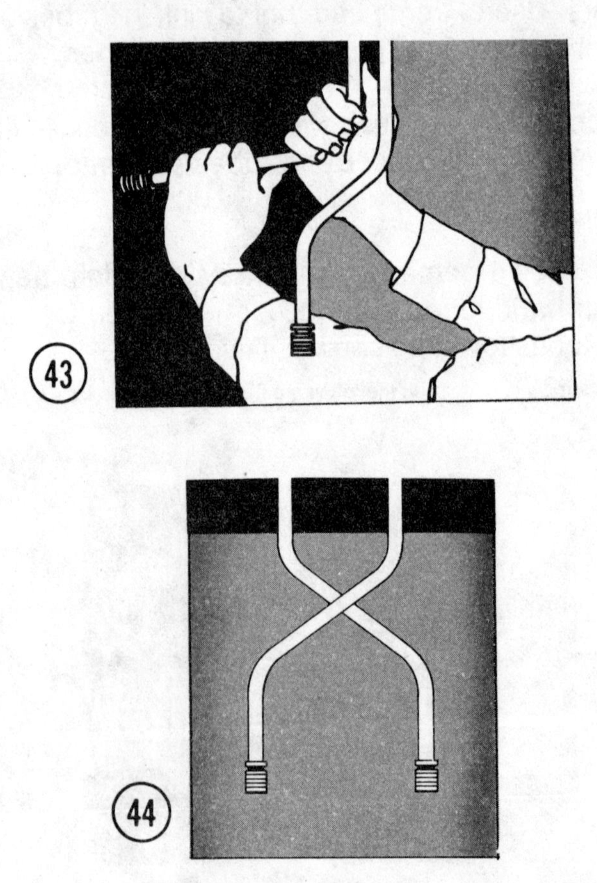

The next step is to apply pipe dope or Teflon tape to all threaded fittings and connect hot and cold water tubes to supply lines. Use one wrench to hold shank of adapter, Illus. 45, another wrench to tighten nut on supply line. If supply line needs to be lengthened, or connecting fittings don't match, the store selling the replacement unit also carries adapter tubes and fittings. Always measure size of existing supply line; always measure distance from end of existing supply line to connection on faucet.

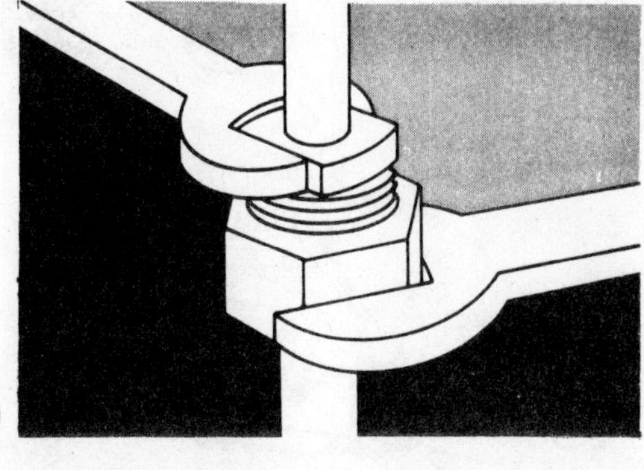

45

After all connections have been made, remove areators from spout and spray, Illus. 46. Flush both hot and cold water lines separately for at least a minute. A small stream of water from a faucet equipped with a spray is normal.

AERATOR PARTS

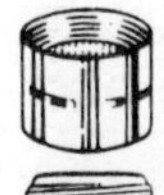

Part No. A-1101. Female Aerator for most faucets with outside threads.

Part No. A-1107. Male Aerator (15/16") faucets w/inside threads. Fit most Am. Std., Briggs, Crane, Delta, Kohler, Repcal, Speakman.

Part No. A-1108. Male Aerator (13/16") faucets w/inside thread. Fit Barnes, Briggs, Chicago, Ind. Brass, Speakman, Koher, others.

Part No. A-1110. Male/female (outside-inside) threaded for use on all faucets with inside or outside threads.

Part No. A-1115. Universal "clip adaptor" aerator for faucet spouts that are not threaded.

Part No. C-3228 (Washer)

Part No. A-1127 (Nylon Screen)

Part No. A-1128 (Perforated Cup)

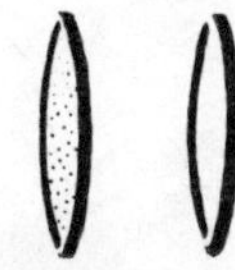

Part No. A-1129 (Copper Screen) Part No. C-3220 (Asbestos Washer)

46

The replacement unit, Illus 36, is easy to service. The manufacturer provides complete and detailed directions. If a leak develops under the handle, shut off water. Loosen set screw, Illus. 47, and lift off handle. Turn water back on. Tighten adjusting ring, Illus. 48, until no water leaks out around stem when faucet is turned on. Replace handle and tighten set screw, Illus. 49. Remove and clean aerator, Illus. 46.

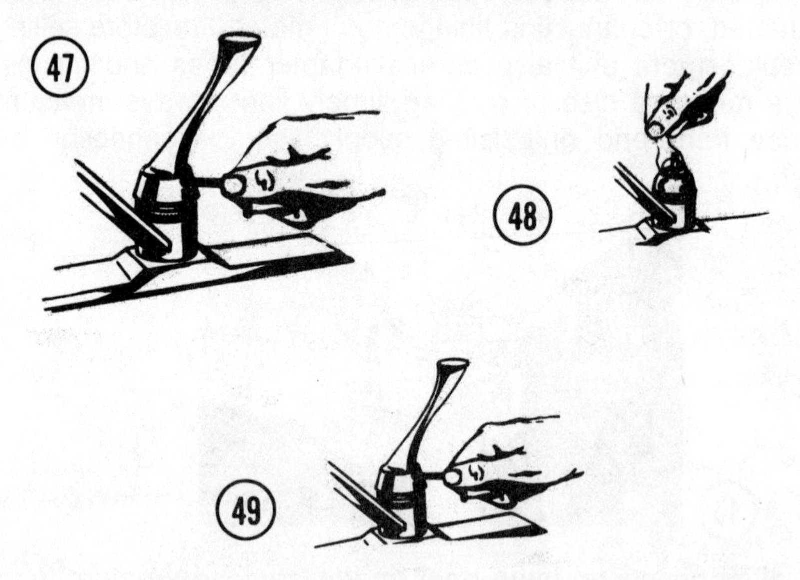

TWO HANDLE KITCHEN FAUCET REPLACEMENT

Two handle replacement faucets, Illus. 29, 50, are available with separate spray, or without spray attachment. Always place black base gasket in position shown. Follow installation previously outlined. Note position of washer and nut. These fasten fixture to sink, Illus. 51.

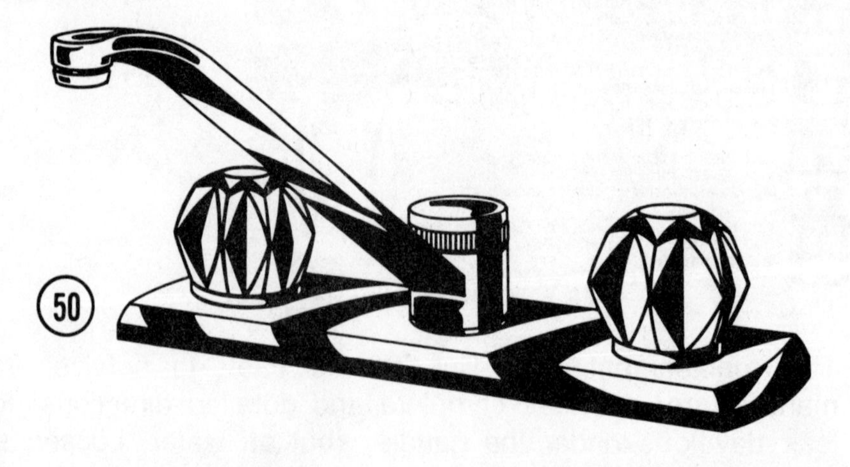

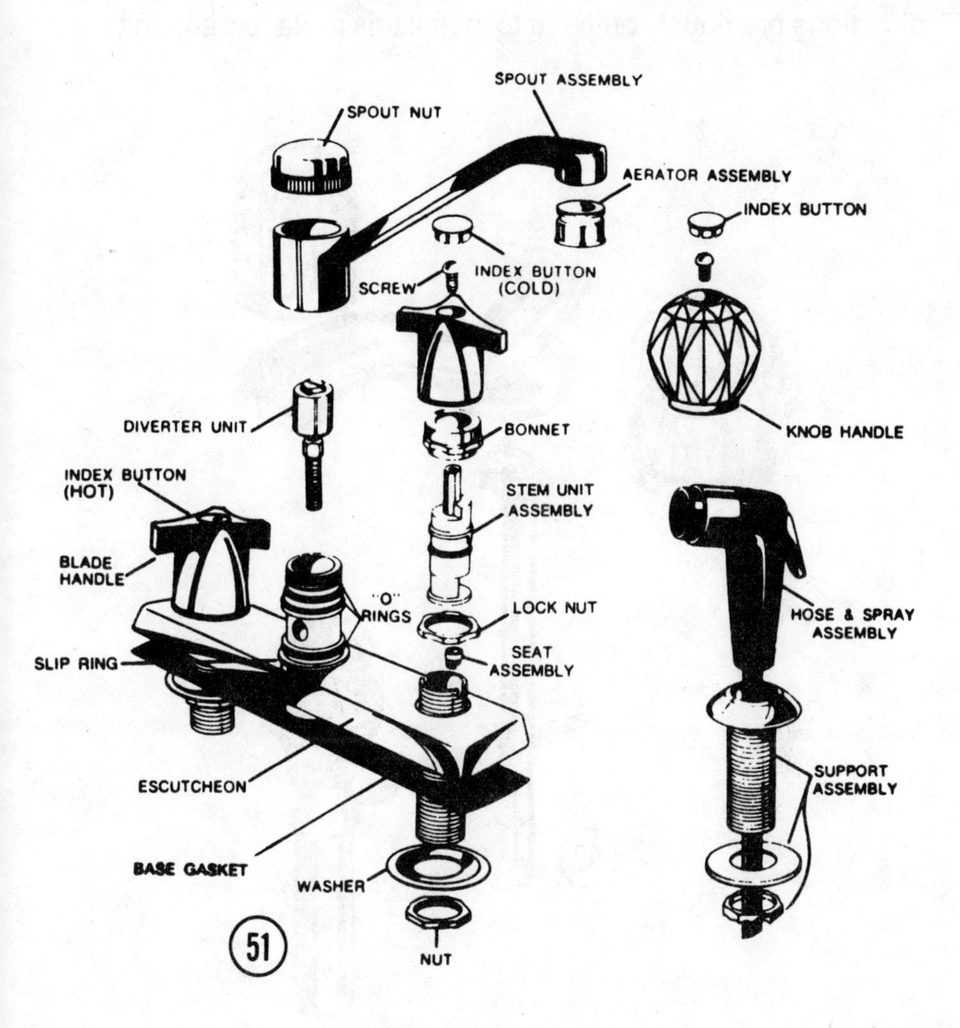
SPOUT ASSEMBLY
SPOUT NUT
AERATOR ASSEMBLY
INDEX BUTTON
INDEX BUTTON
(COLD)
SCREW
DIVERTER UNIT
BONNET
KNOB HANDLE
INDEX BUTTON
(HOT)
STEM UNIT
ASSEMBLY
BLADE
HANDLE
"O"
RINGS
LOCK NUT
HOSE & SPRAY
ASSEMBLY
SLIP RING
SEAT
ASSEMBLY
SUPPORT
ASSEMBLY
ESCUTCHEON
BASE GASKET
WASHER
NUT
51

TWO HANDLE WIDE SPREAD LAVATORY FAUCET REPLACEMENT

Illus. 52 shows a popular wide spread lavatory replacement faucet. Illus. 53 shows installed position of each part. Follow directions previously outlined to install this replacement unit.

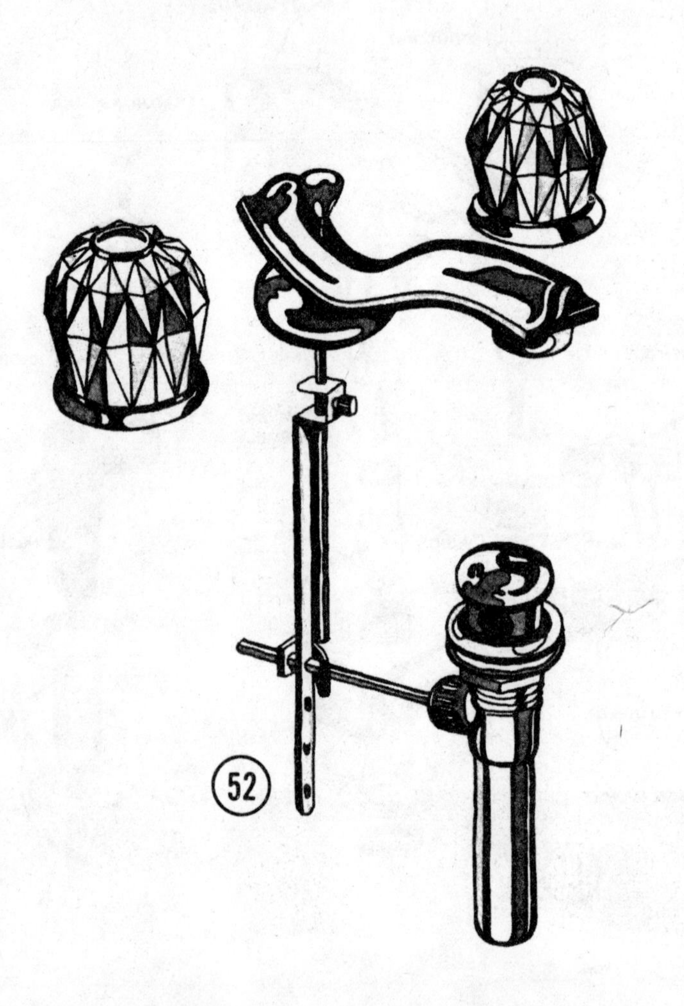

Remove spout from box. Press black base gasket, Illus. 53, 54, in center hole. Place spout in hole. Working from below, place gasket, washer on stem, Illus. 55. Fasten spout in position. Apply pipe dope or Teflon tape to threads on ends of flexible hoses and fasten in position.

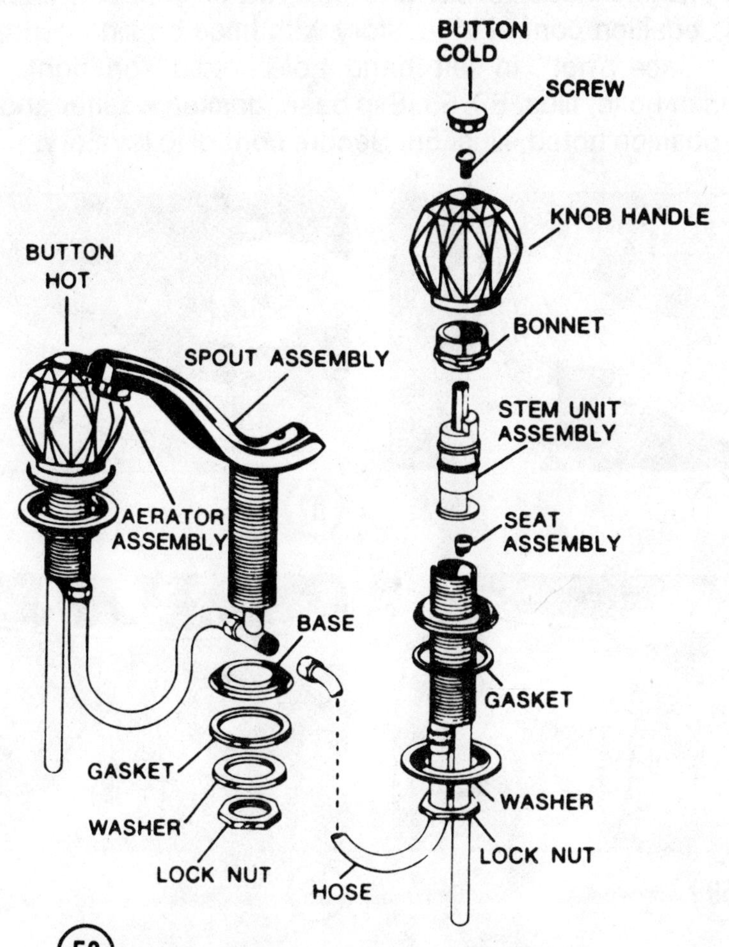

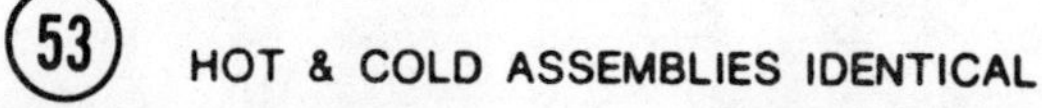

HOT & COLD ASSEMBLIES IDENTICAL

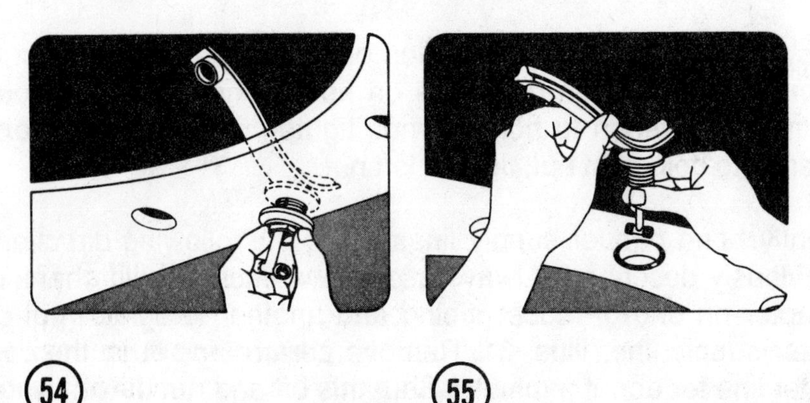

Attach flexible hose to hot and cold water controls, Illus. 56. Always position control in lavatory with hose on side nearest to spout. Place "hot" in left hand hole, "cold" on right. Note position of hose, Illus. 57, 58. Slip base, gasket, washer and lock nut, in position noted, Illus. 58. Secure control to lavatory.

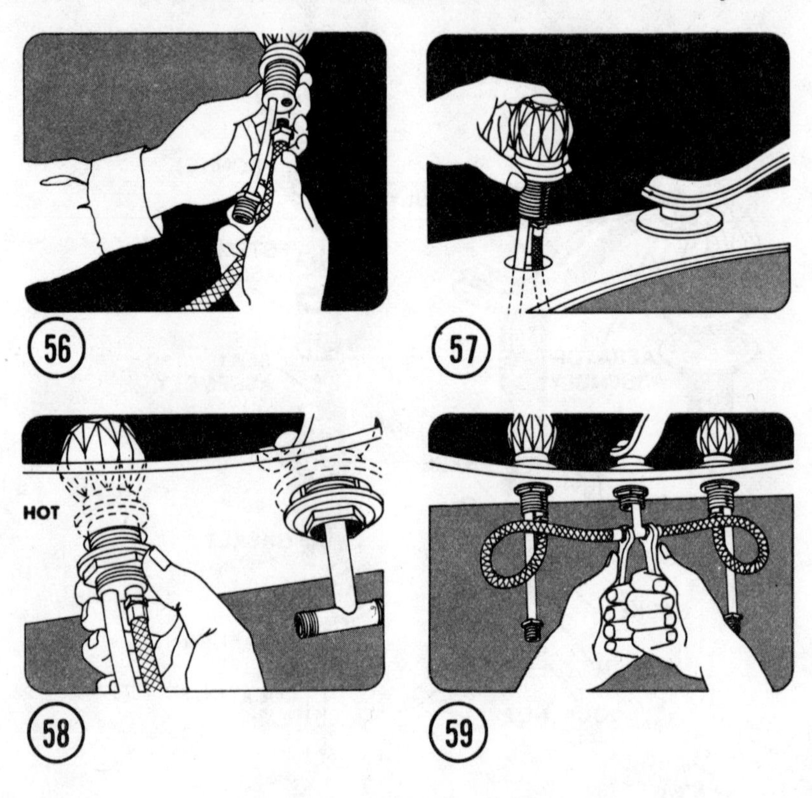

56

57

58

59

Loop hose, Illus. 59, apply Teflon or tape dope and connect to spout. Finger tighten both nuts on hose connections to spout. Using one wrench to hold left nut, tighten right nut. Using one wrench to hold right nut, tighten left nut.

Gently bend copper supply lines, Illus. 60, following directions previously described. Always use one wrench to hold shank of adapter on end of faucet tubing and another to tighten nut on water supply line, Illus. 61. Remove aerator and flush the cold water line for about a minute. Shut this off and run the hot water line. Replace aerator.

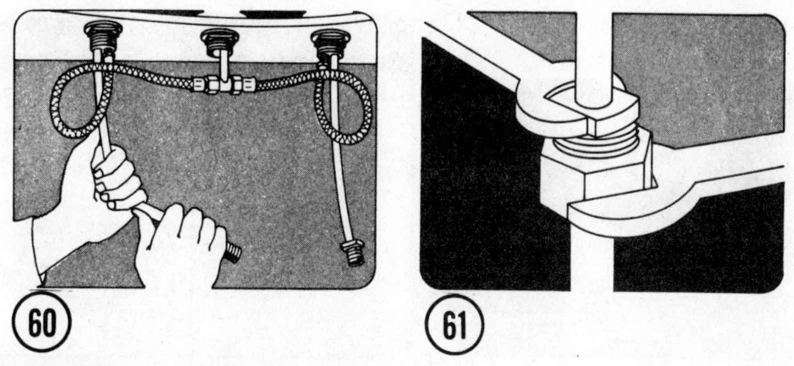

(60) (61)

TO INSTALL A POP-DRAIN

Unscrew tail piece A, Illus. 62, 63.

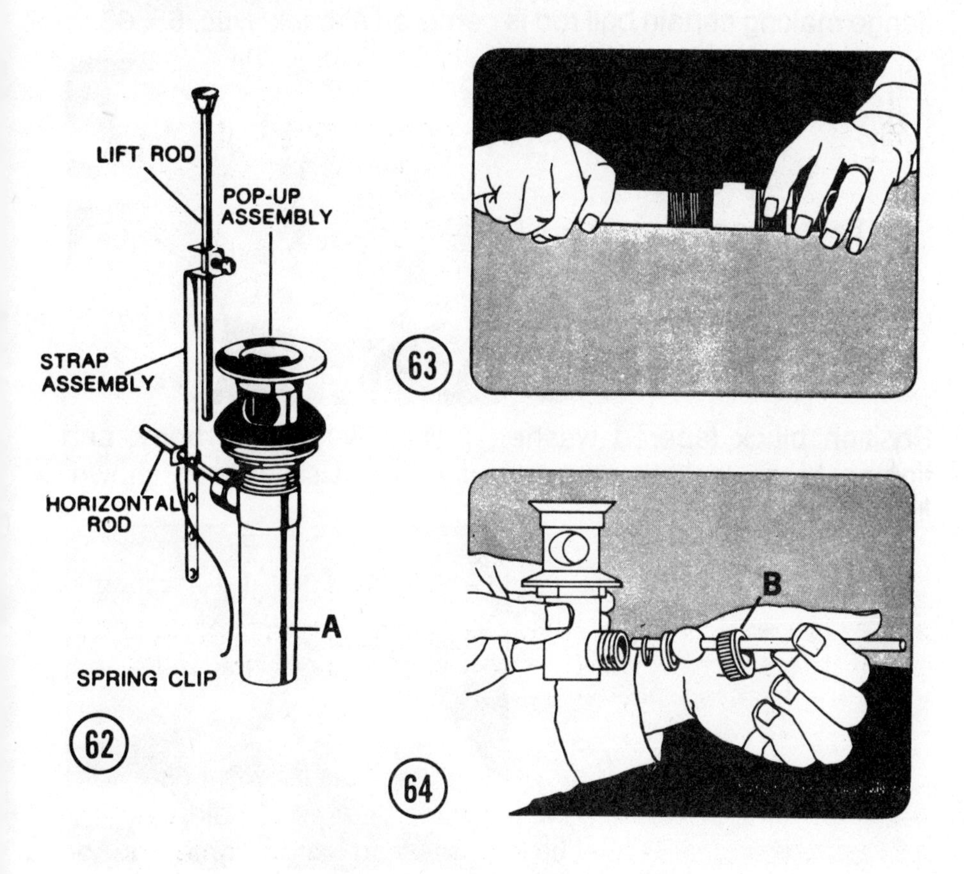

(62) (63) (64)

Remove nut B off side. Slip black gasket and white seat on ball rod. Slide ball into position and fasten nut finger tight, Illus. 64.

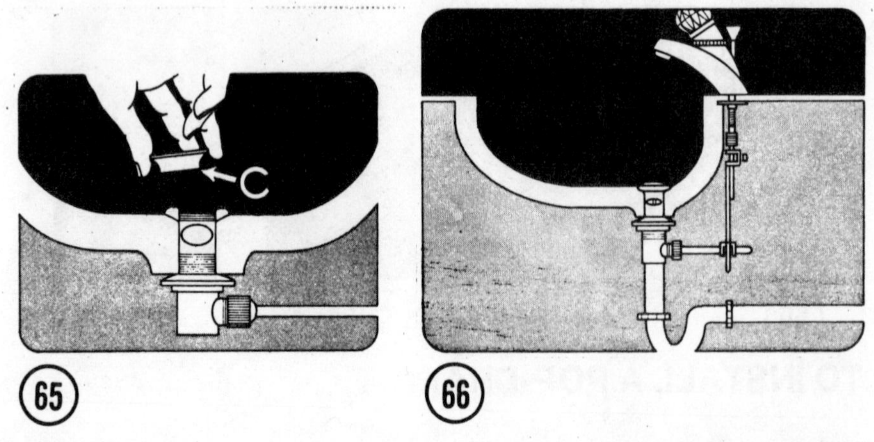

Unscrew flange C. Lay a ring of putty around drain hole. Tighten flange making certain ball rod is centered at back, Illus. 65, 66.

Excess putty can be removed after flange seats it self.

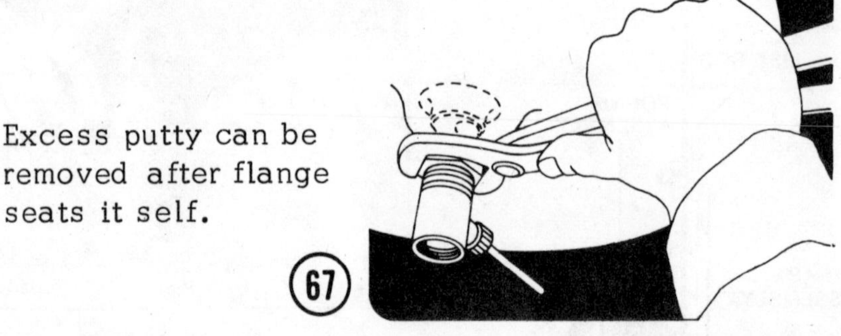

Position black tapered washer, friction washer on body, and tighten brass locknut on body, Illus. 67. Use wrench shown, Illus. 10.

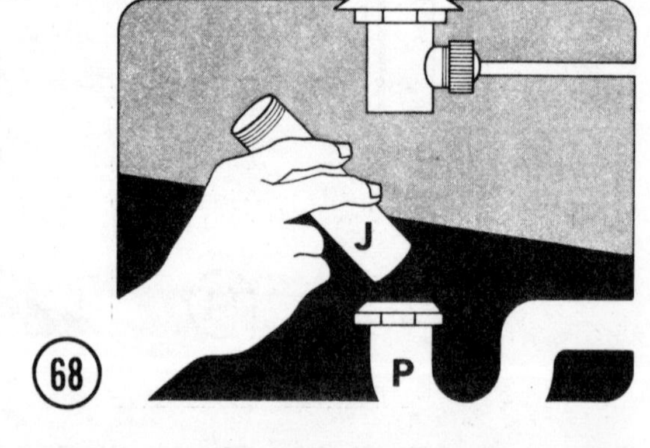

Replace washer in nut on P trap, Illus. 68, if same is needed. Slide tailpiece in P trap and screw tail into body. Tighten trap nut.

Insert lift rod through hole in back of spout, Illus. 69. Place clip, Illus. 70, on rod. Slip punched strap on rod and adjust position of clip so it opens and closes drain, Illus. 71.

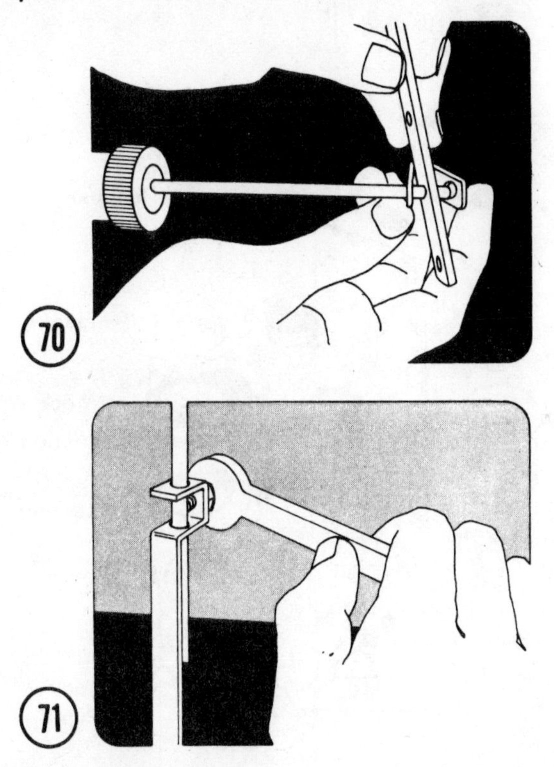

After completing installation, remove aerator and flush both lines, then replace aerator.

Illus. 72 shows position of parts in many lavatory faucets. The threaded shank of fixture A is inserted through opening in lavatory. Shank washer B is placed snug against bottom surface. Locknut C holds fixture in position.

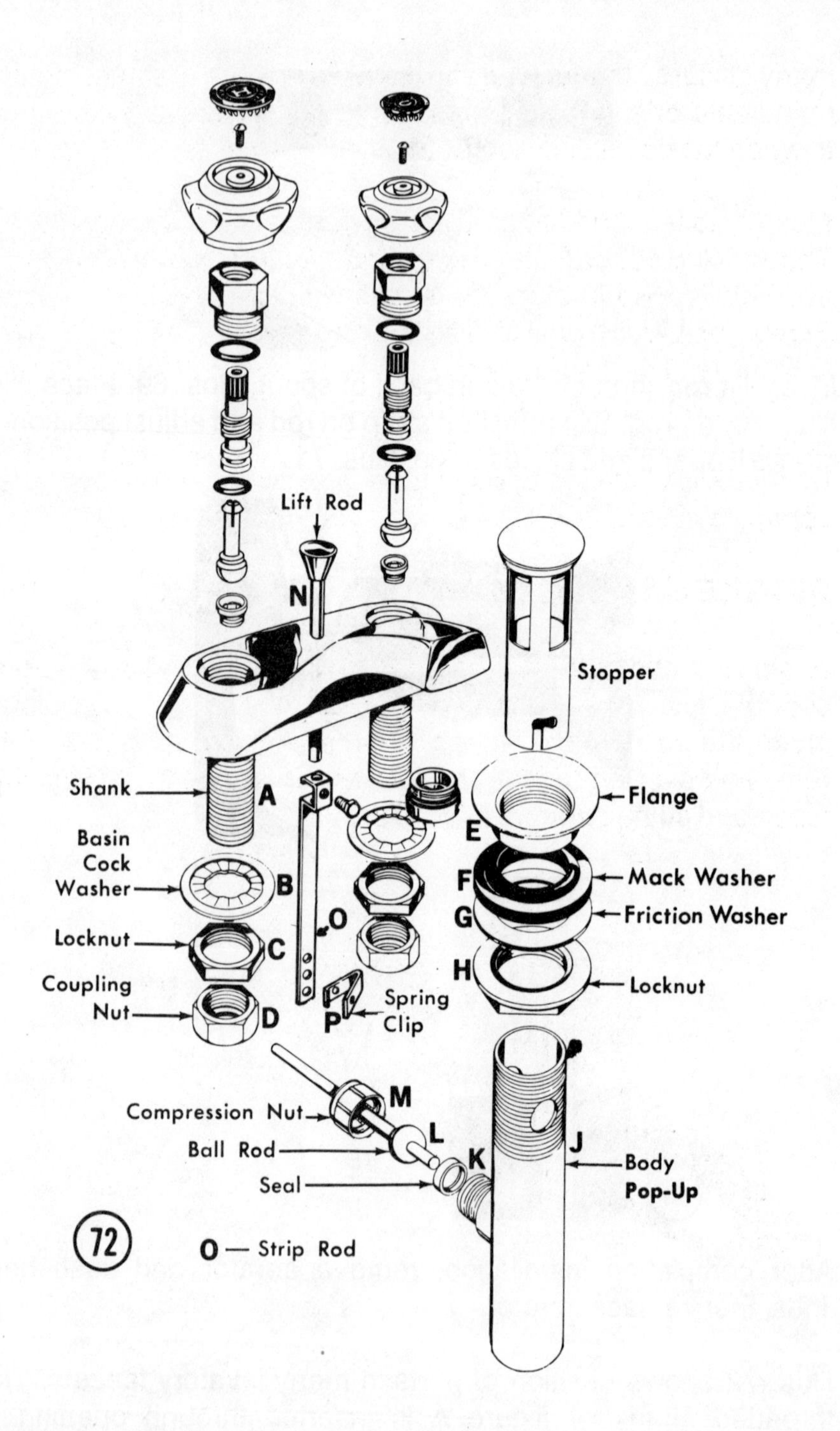

Coupling nut D slips on supply line. The line is then flanged and fastened to A. While some connections are made with a washer, many fixture manufacturers have eliminated same. Follow manufacturer's, or plumbing supply dealer recommendations as to when to use sealing compound and/or washer.

Most installers bed flange E for waste line in setting compound. The mack washer F, is placed under fixture with beveled face up. This is held in place by friction washer G and a locknut H. Screw locknut H onto drain J. Screw drain J into E. Tighten locknut H. The pop-up waste lever N slips into rod O and is fastened with a setscrew. The ball rod L is fastened to O with a spring clamp P. Coupling M fastens L to J. Be sure beveled washer K is placed in position to receive ball on rod. Waste line J connects to trap, Illus. 68.

SINGLE HANDLE LAVATORY FAUCETS

A single handle lavatory faucet replacement without pop-up control, Illus. 73, 74; or with pop-up drain, Illus 75, is installed in much the same way as a single handle faucet for a sink. After removing existing faucet, the new one is placed in position with tubes and studs through holes, Illus. 76.

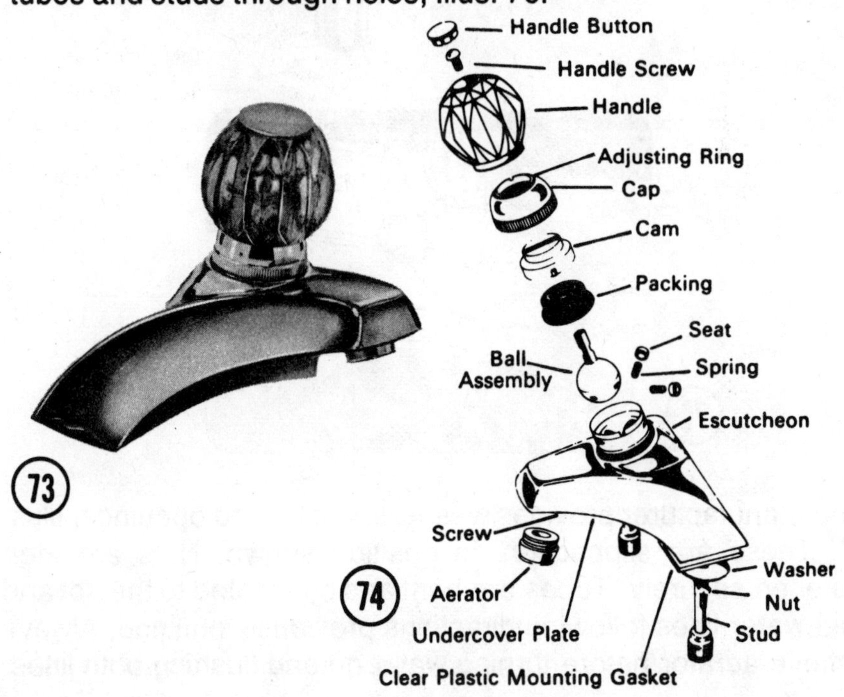

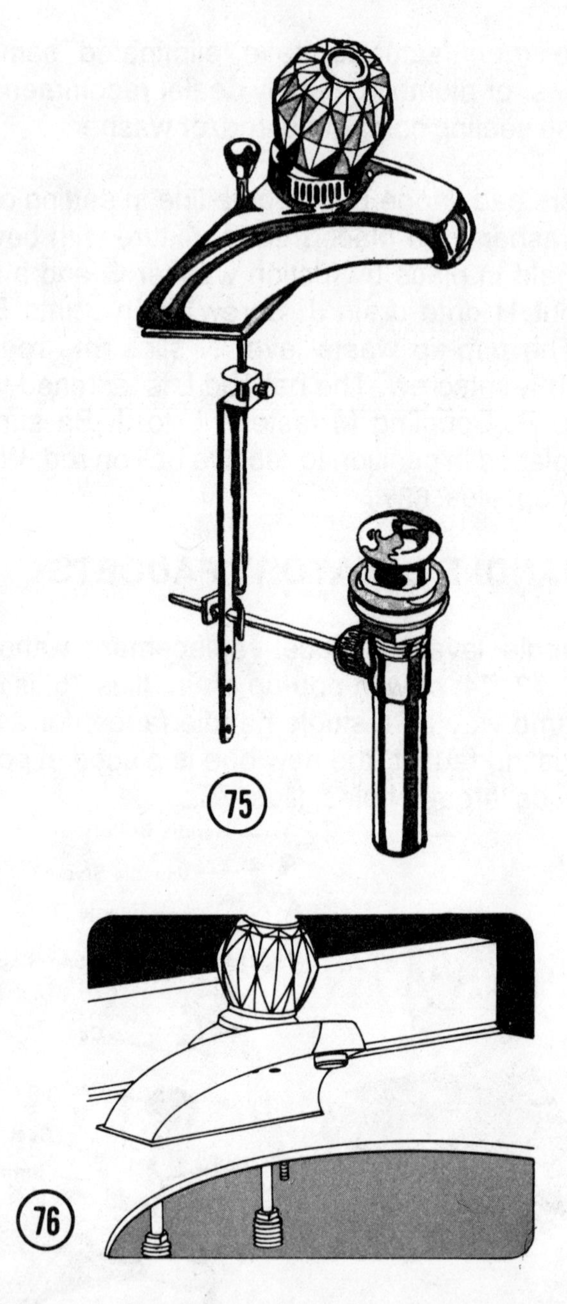

One manufacturer provides washers with slotted openings, Illus. 77. These are slipped on, in position shown. Nuts are then fastened securely. Tubes are bent and connected to the hot and cold water lines following directions previously outlined. Always remove aerator before turning water on and flushing both lines.

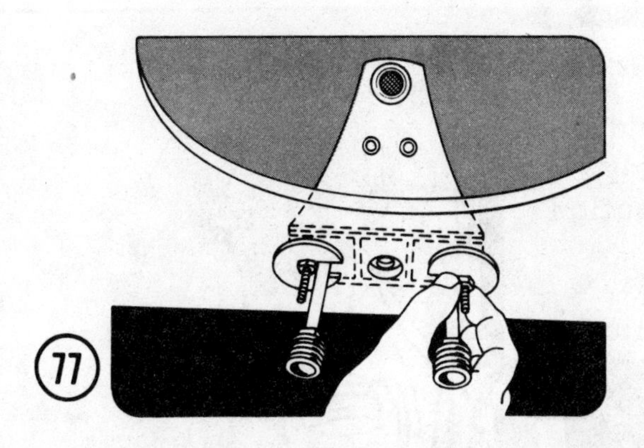

If existing faucet contains a separate control for the pop-up drain, purchase the replacement shown in Illus. 62.

TUB AND SHOWER CONTROL REPLACEMENT

To replace shower and tub controls, Illus. 78, shut off water, open control and remove handle, escutcheon plate and faucet. A replacement unit contains components shown in Illus. 79. Note position of each part, Illus. 80.

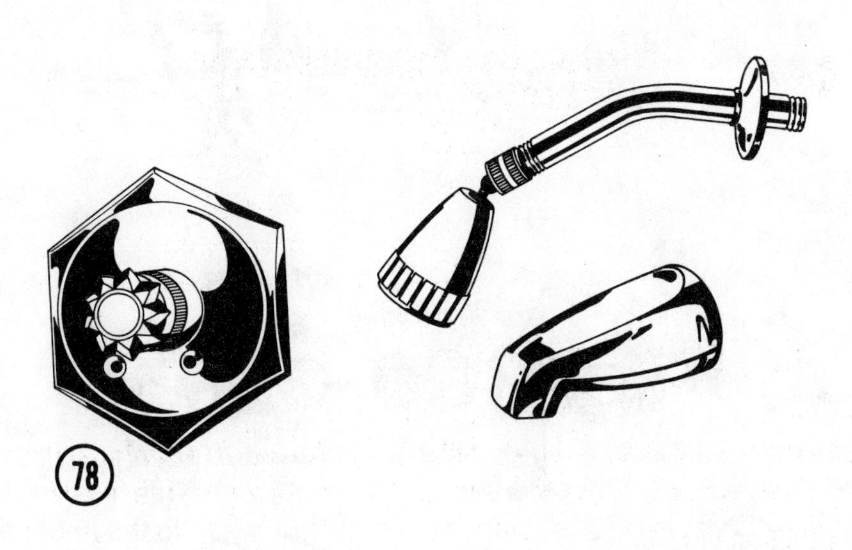

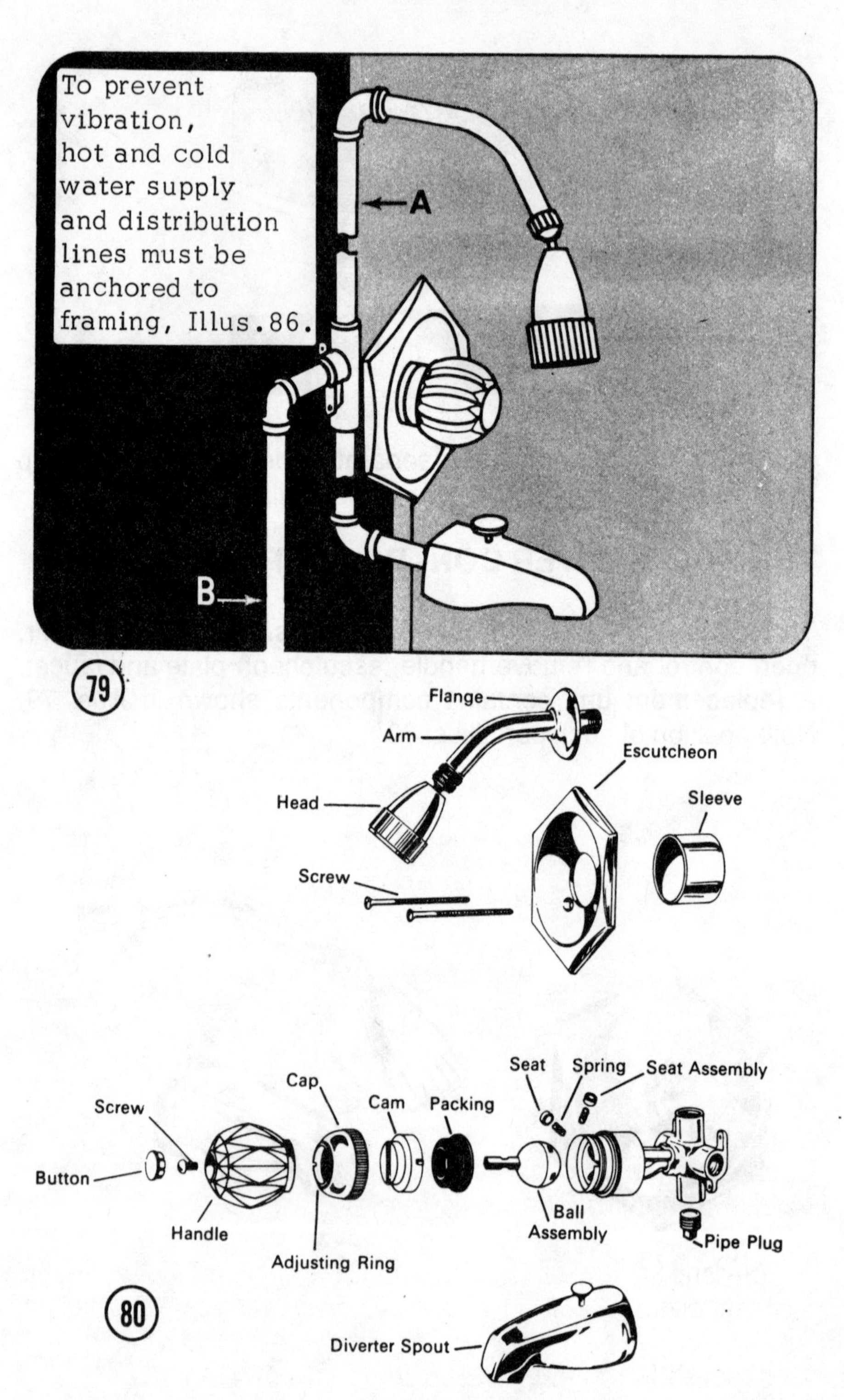
To prevent vibration, hot and cold water supply and distribution lines must be anchored to framing, Illus.86.
A
B
79
Flange
Arm
Escutcheon
Head
Sleeve
Screw
Seat
Spring
Seat Assembly
Cap
Cam
Packing
Screw
Button
Handle
Adjusting Ring
Ball Assembly
Pipe Plug
80
Diverter Spout

Install main valve, Illus. 81. Valve is marked "up and down". Note and follow directions on valve.

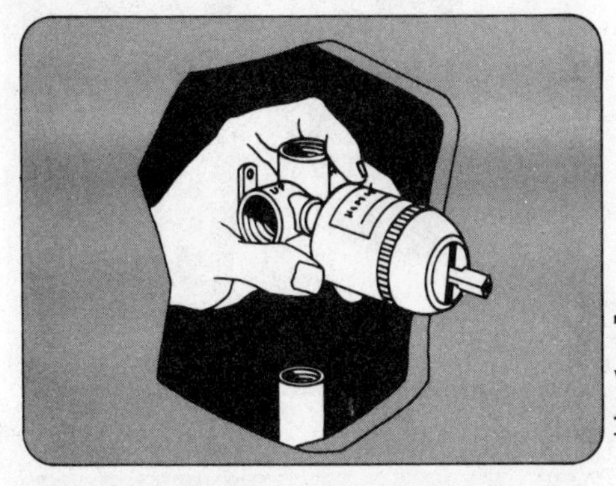

81

This mixing valve mounts from front.

If your present supply line only feeds the tub, and you want to install a shower, it will be necessary to open the wall to install the extra length of pipe A, Illus. 79.

If you are only making a tub installation, it will be necessary to plug top outlet, Illus. 82.

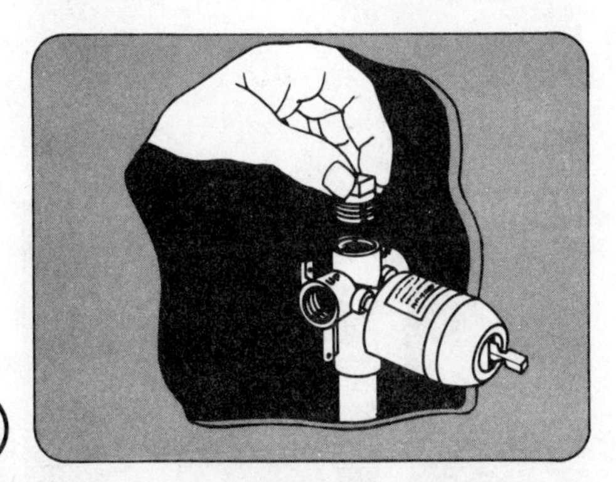

82

If you are only making a shower installation, no tub control, plug the bottom hole, Illus. 83. Manufacturer of replacement controls supplies a plug. Always apply pipe dope or Teflon tape to all threaded fittings before making a connection.

If tub and shower combination installation is being made, connect diverter spout to a new 1/2" threaded nipple B, Illus. 84.

Connect shower head, Illus. 85.

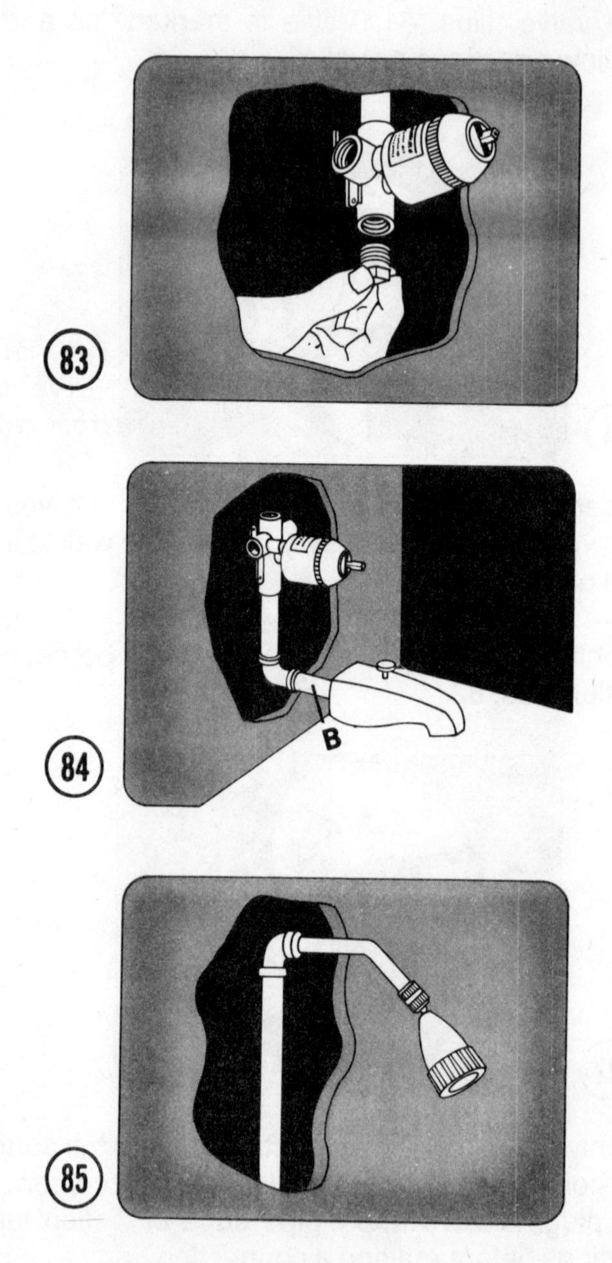

If you are making a new installation, frame opening for water supply lines in position shown, Illus. 86. Book #682, How To Install An Extra Bathroom, contains detailed step-by-step directions that explain how to rough in supply and waste lines, install fixtures and frame in a bathroom.

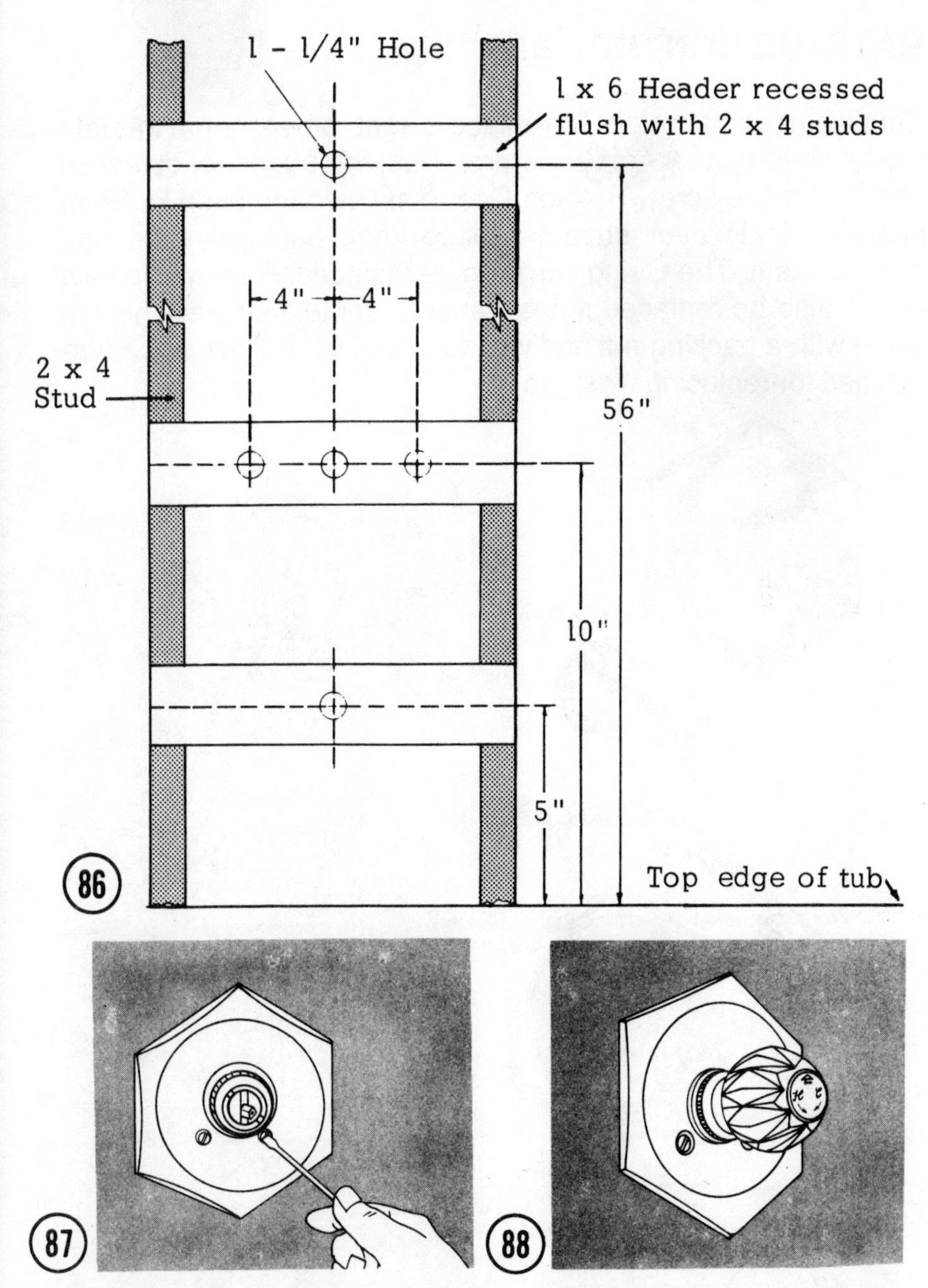

Unscrew shower head before turning water on. Flush both hot and cold water lines for about a minute. This will wash out all particles that might have been picked up during assembly. Remove handle and fasten escutcheon plate in position, Illus. 87. Replace handle, Illus. 88.

BATHTUB CONTROL REPAIRS

Shut water off at meter. To replace a seat, or washer in this late model bathtub and shower control, Illus. 89, 90, pry or screw off cap A, remove screw B, knob C, and escutcheon plate D. Stem bushing E fits over stem F. Place knob back on stem and remove stem. The O ring can then be replaced. Renewable seat G can also be replaced if need arises. Some stems are held in place with a packing nut and washer, Illus. 91. Follow procedure outlined for replacing washers.

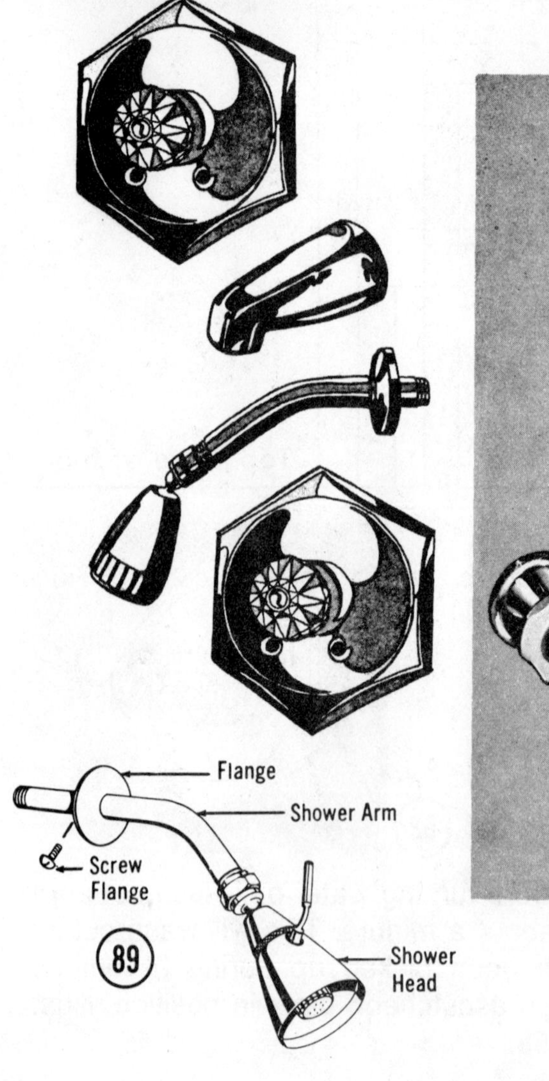

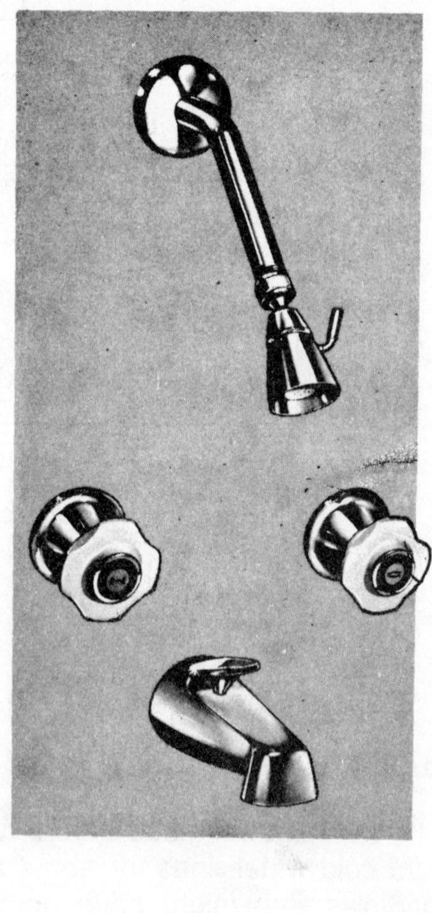

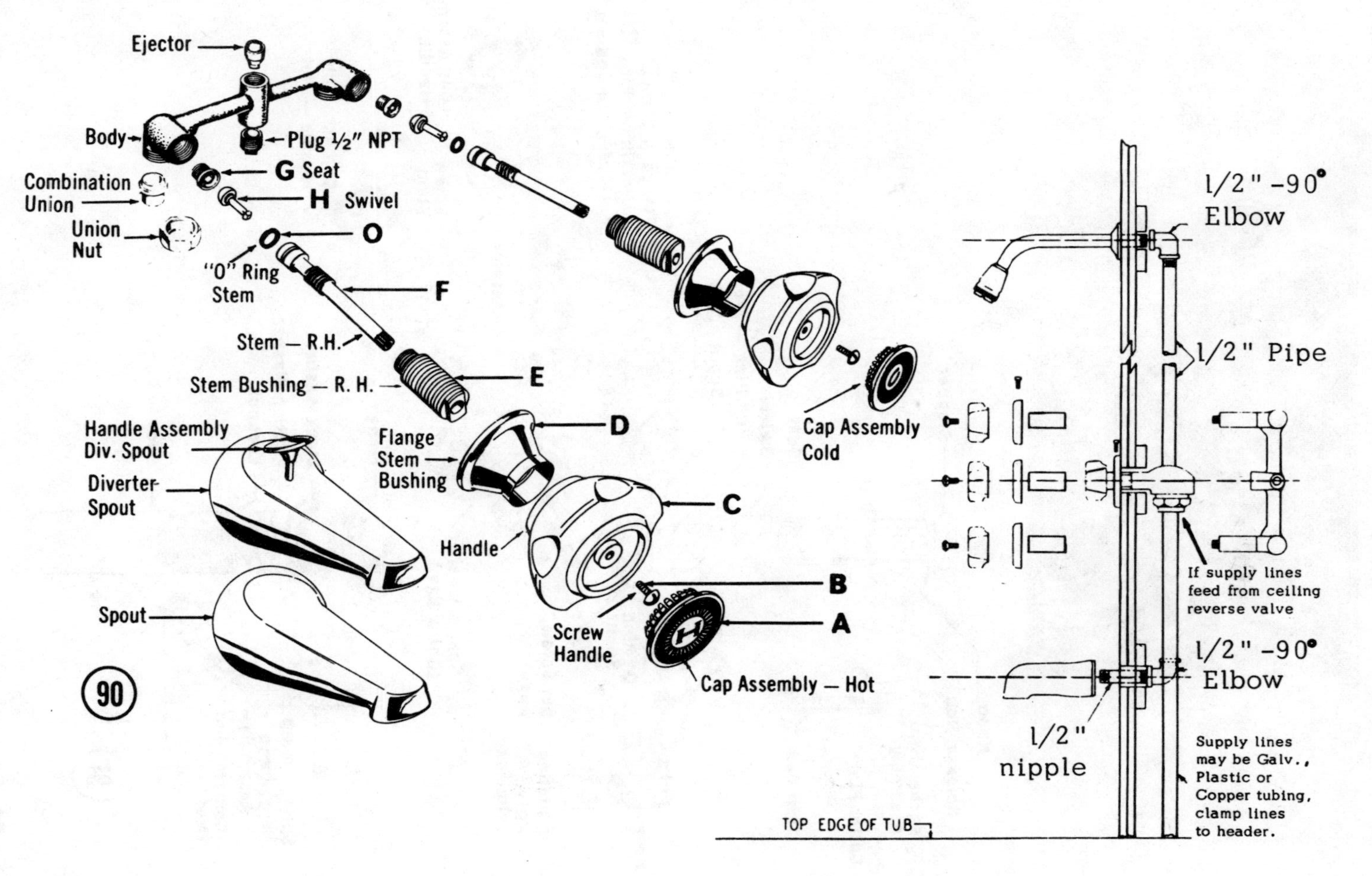
Ejector
Body
Plug ½" NPT
G Seat
Combination Union
H Swivel
Union Nut
O
"O" Ring Stem
F
Stem — R.H.
Stem Bushing — R. H.
E
Handle Assembly Div. Spout
Flange Stem Bushing
D
Diverter Spout
C
Handle
B
Spout
Screw Handle
A
Cap Assembly — Hot
Cap Assembly Cold
90
l/2" -90° Elbow
l/2" Pipe
If supply lines feed from ceiling reverse valve
l/2" -90° Elbow
l/2" nipple
Supply lines may be Galv., Plastic or Copper tubing, clamp lines to header.
TOP EDGE OF TUB

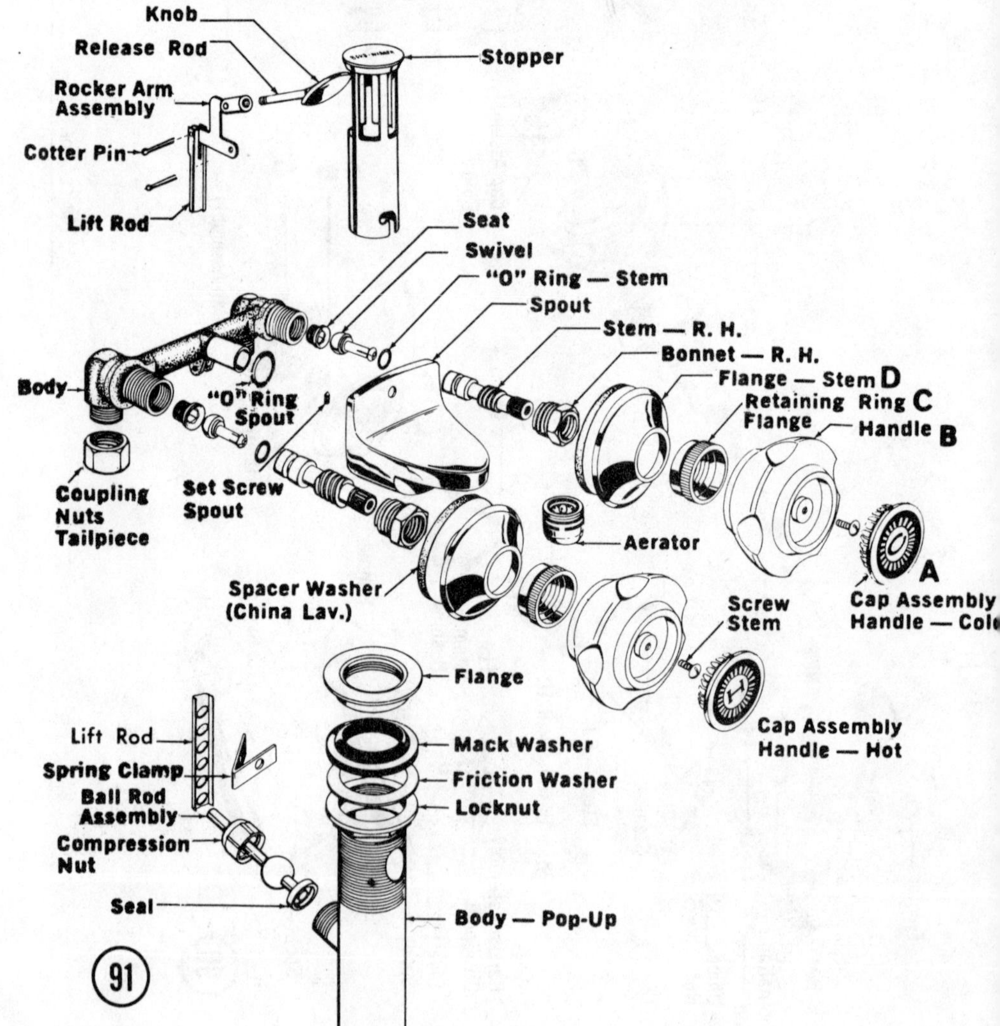
Knob
Release Rod
Stopper
Rocker Arm Assembly
Cotter Pin
Lift Rod
Seat
Swivel
"O" Ring — Stem
Spout
Stem — R. H.
Bonnet — R. H.
Flange — Stem D
Retaining Ring C
Flange
Handle B
Body
"O" Ring Spout
Coupling Nuts Tailpiece
Set Screw Spout
Aerator
A
Spacer Washer (China Lav.)
Screw Stem
Cap Assembly Handle — Col
Flange
Cap Assembly Handle — Hot
Lift Rod
Mack Washer
Spring Clamp
Ball Rod Assembly
Friction Washer
Locknut
Compression Nut
Seal
Body — Pop-Up
91

BATHTUB DRAIN

You can remove stoppage from some bathtub drains by removing drain plate A, Illus. 92, 93. The plunger assembly B can be removed by loosening screws C. Lift plate and plunger out of pipe. A small snake can be worked through drain and trap. Most reliable builders provide an access panel that permits working on bathtub drain and trap. This is usually screwed or hinged on other side of wall on drainage end of tub. The tub trap can be serviced following same procedure for a sink trap.

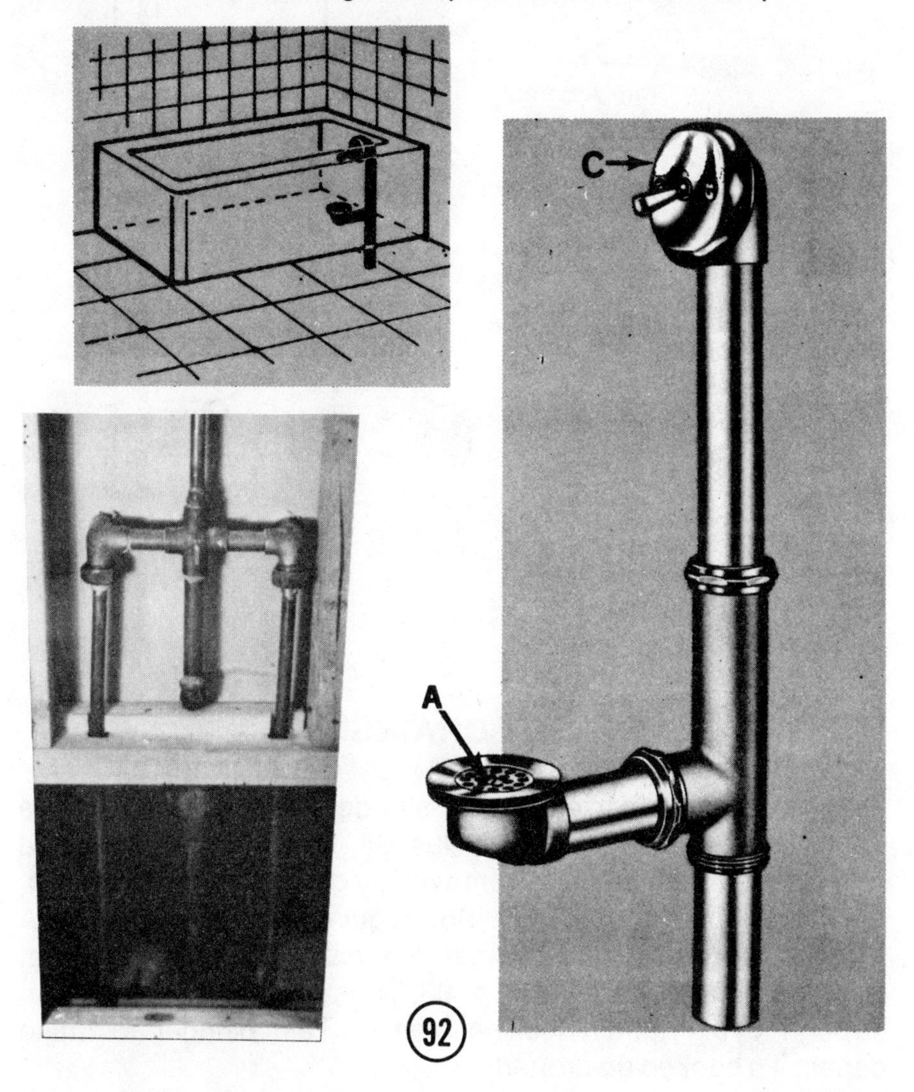

92

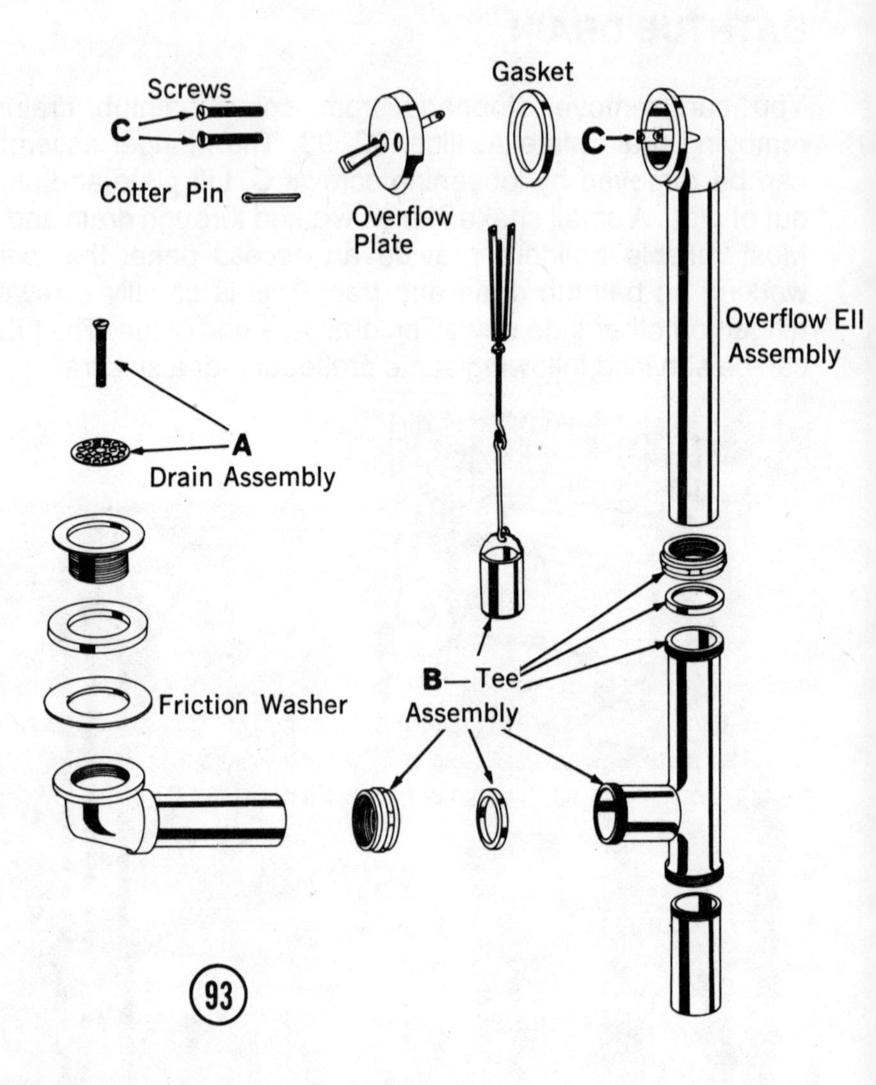

CLOGGED LAVATORY DRAINS

Hair, odds and other ends, that slip down a lavatory drain are usually caught by the stopper, Illus. 94. This should be removed and cleaned. Some can be removed by giving the top a half turn, then pull up. Others, Illus. 95, require unscrewing nut at A and pulling rod. Lift out plunger. Remove accumulated waste. Replace plunger so eye, Illus. 96, lines up with rod. Insert rod through eye. Tighten locking sleeve. Test plunger to make certain it's hooked up properly.

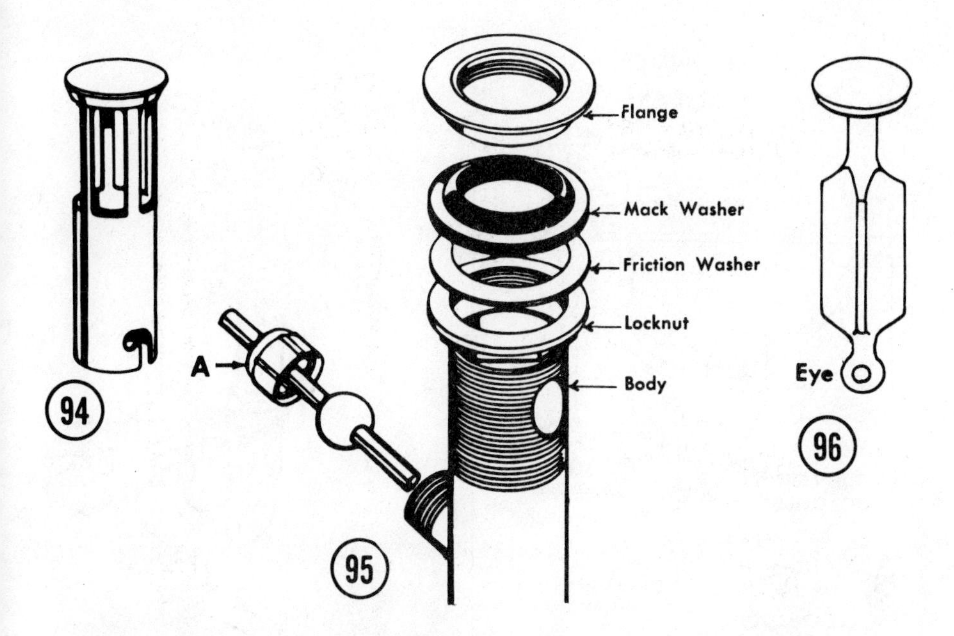

If the stopper doesn't free drain, the problem is probably in a clogged trap. First try a drain cleaner following manufacturer's directions. If this doesn't free up fixture, remove plunger and try a small snake, Illus. 97. Carefully turn handle on snake to turn head. You can work this kind of snake clear through trap.

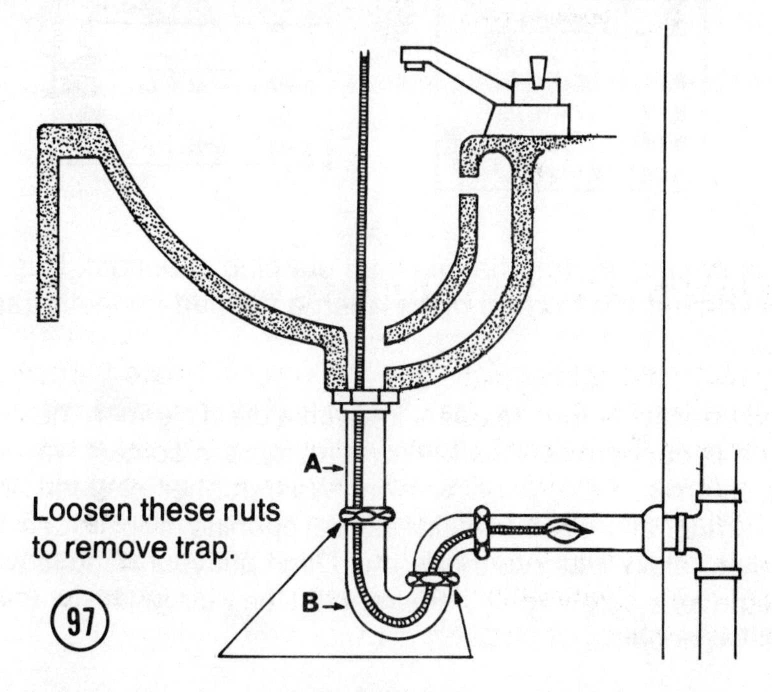

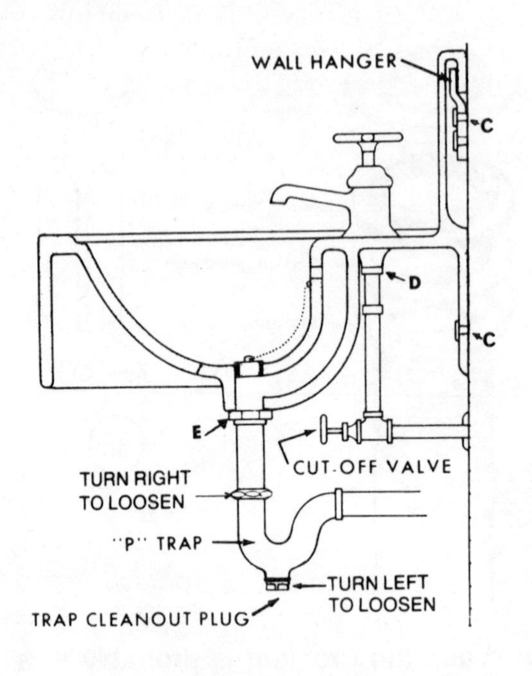

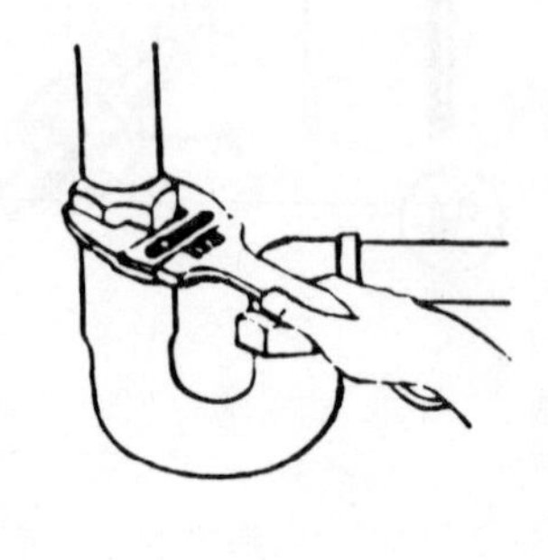

TRAP SWIVEL-JOINT WASHERS

Part No.	Description
W-810	2¼"x1 15/16"x3/32"
W-811	1 15/32"x1¼"x1/16"
W-828	1 27/32"x1¼"x3/32"
W-828A	1 11/16"x1 7/16"x3/64"
W-829	1⅜"x1⅛"x3/64"

TRAP WASHERS

Part No.	Description
W-833	1 7/16"x1¼"x3/32" Rubber
W-836	1 21/32"x1 13/64"x5/16" Rbbr.
W-849	1 7/16"x1 3/16"x3/32" Fibre

If sink or lavatory trap has a clean out plug at bottom, Illus.97 , remove plug. Be sure to place a large flat pan beneath trap to catch waste.

If you previously tried to open line with a drain cleaner, the water in trap is dangerous. Don't allow it to come in contact with your hands, arms, face, eyes or skin. Wear rubber gloves. Cover wrists and all exposed skin before opening trap. Open trap slowly to allow water to trickle out. Don't put your hand in water to retrieve a screw, plug, or wrench. Don't throw this water on the lawn or plants.

If there's no clean out plug at bottom of trap, and you can't get a snake through trap, (this is a very unlikely situation), remove trap completely by loosening packing nuts B, Illus. 97.

To loosen, turn packing nut to right. When reassembling, insert A into B as far as it will go, then start fastening packing nut. Always replace washers rather than reuse old ones. Buy the same kind and size. If they have a beveled face, be sure to replace in exact position.

CLOGGED SINK DRAIN

Before leaving a house for three or more days, flush sink with lots and lots of hot water. This will loosen up and wash down particles of food, grease, etc. Then, to play safe, use a good drain cleaner. But make certain all cleaner is flushed out of trap and lines before departure. Many of even the best drain cleaners harden up when left longer then directions specify, so use plenty of hot water.

If your sink does clog, use care in selecting a drain cleaner. Most contain dangerous chemicals. Read and follow directions on can. Never allow water or any dampness to get into container. If you put a cap on a can containing only a few drops of water, it could blow up in your face.

If a plunger, drain cleaner, or a snake hasn't opened line, the next step is to open up the nearest drain plug, Illus. 8. Remove plug, run snake through line. If plug is rusted, use penetrating oil as suggested previously. Give oil time to penetrate, then tap lightly with a hammer to jar rust.

Always open cleanout plug nearest fixture. If the first cleanout plug doesn't open the line, try the next one, Illus. 8.

Always use an electric heater or heat lamp to warm up a trap. Traps, insulated against room heat by under sink enclosures, frequently clog in severe cold weather. In cold weather always keep door of a sink enclosure open.

Never use drain cleaner if sink is equipped with a garbage disposal unit. Use snake through trap. A pound of washing soda used weekly will keep most disposal drains open.

PREVENTATIVE MAINTENANCE

Intelligent car owners know the value of keeping a car serviced, yet relatively few show the same concern for the plumbing in their homes. Neglect frequently costs big money. Soap, hair and bits of waste tend to coat drainage lines. This attracts minute particles of hair, even those from an electric razor. In time, the opening narrows and clogs. Gases backup into your house. Before you fully realize what has happened, you experience a blockage that resists lye, crystals, solvent or other drain cleaners.

To free a lavatory waste line that's frequently connected to a bathtub, use an electric snake, Illus. 12. This necessitates removing overflow valve C, Illus. 93. Use a ¼" electric snake with a ¼" flexible auger. Do not use a larger auger or snake. Insert down overflow pipe at C, Illus. 93.

This is a two man job. One man with gloves must guide the snake while the second man operates the machine. If you can't penetrate blockage with snake in forward gear, try reverse.

If your bathroom begins to develop a nauseous odor that seems to resist all cleaners, try this. Plumbing supply houses sell Cloroben PT. While primarily a drain cleaner, it's especially good in removing odor. Mix with water exactly as directions on can specify. Pour down bathtub drain. Immediately clean the chrome and enamel with a very wet rag. Wear rubber gloves. Allow mixture to set following manufacturer's directions, then flush with fresh water. Don't get smart fast and allow it to set longer than directions recommend. Use on a regular basis and the results are truly worth the effort.

HOW TO INSTALL A LAVATORY

If fixture is beyond repair and needs replacement, buy an exact size replacement. Shut water off at shutoff valve. Place open pan below and disconnect water line at D, Illus. 97. Disconnect waste line at E, loosen nut. Lavatory is fastened to wall at C.

Lavatories can be fastened to the wall by hanging top edge over wall bracket C, Illus. 97, and/or, through fixture at lower C, or to legs. If you want to hang it to the wall, use brackets that come with lavatory. Always install a replacement lavatory at height that permits connecting present water and waste lines.

TOILET REPAIRS, REPLACEMENT

If a toilet begins to clog, or slow down carrying waste out of the bowl, the trouble might be traced to a change in toilet paper, Kleenex, etc. Soft tissues frequently stick to sides of a bowl, trap, or waste line, while coarse paper flushes down.

First try a plunger. If this doesn't clear drain, use a short snake in bowl, Illus. 13. If this doesn't clear obstruction, locate nearest cleanout plug in main waste line, and use a long snake, Illus. 8, 12.

Always identify whether a real old toilet is one piece or two piece. Also measure spacing of seat bolts, center to center. Do they measure 5½", 7", 7½" apart? This can be an important clue to its actual model number.

Note whether seat bolt (frequently called seat post) goes through front of water tank, or into top of bowl. Also note whether flush handle is on front of water tank or on side.

Note whether bowl is fastened to floor with four bolts or two. The china knobs, Illus. 98, covering nuts, or covering plates, Illus. 99, indicate how many bolts hold toilet to flange.

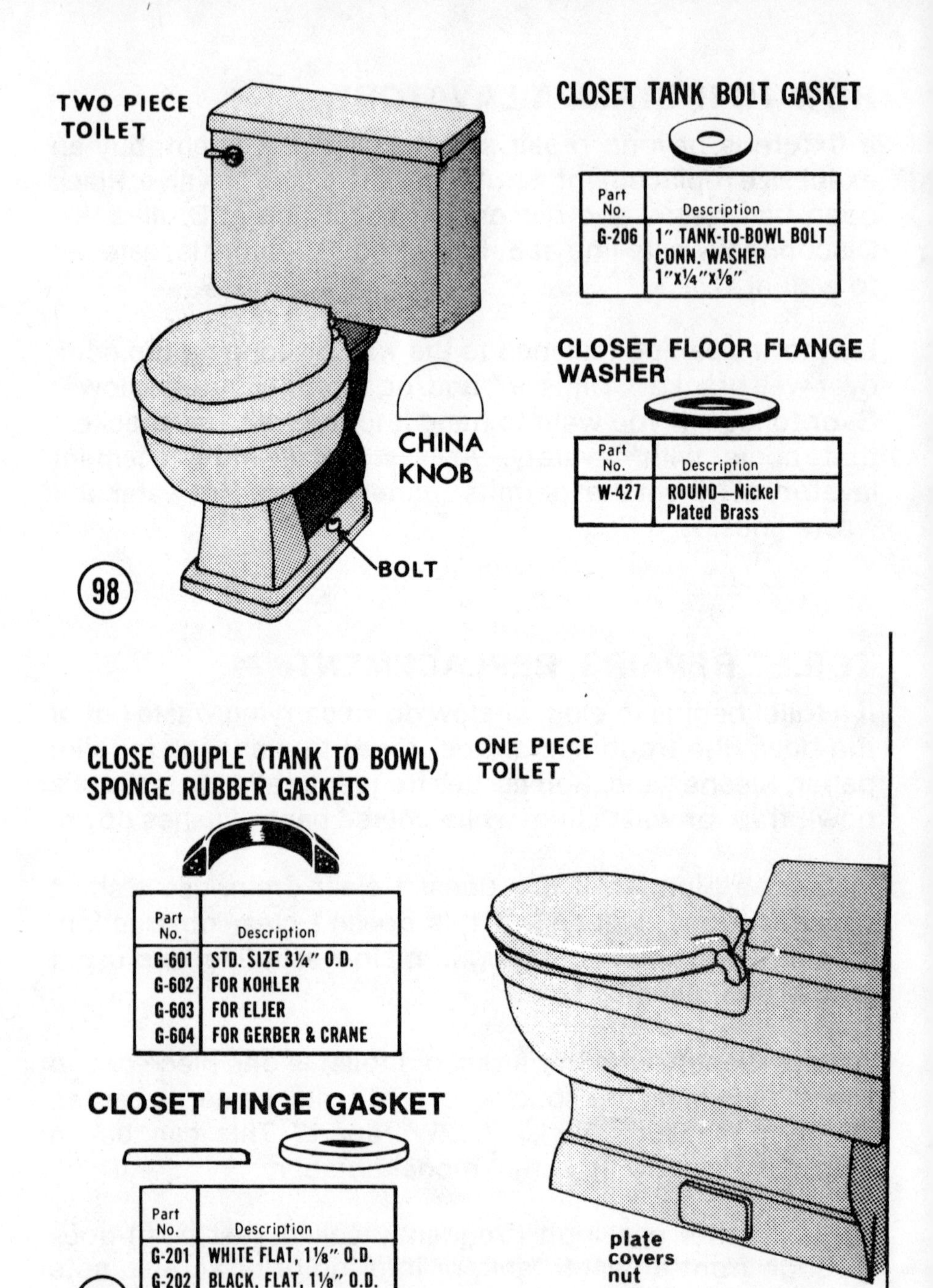

CLOSET TANK BOLT GASKET

Part No.	Description
G-206	1″ TANK-TO-BOWL BOLT CONN. WASHER 1″x¼″x⅛″

CLOSET FLOOR FLANGE WASHER

Part No.	Description
W-427	ROUND—Nickel Plated Brass

CLOSE COUPLE (TANK TO BOWL) SPONGE RUBBER GASKETS

Part No.	Description
G-601	STD. SIZE 3¼″ O.D.
G-602	FOR KOHLER
G-603	FOR ELJER
G-604	FOR GERBER & CRANE

CLOSET HINGE GASKET

Part No.	Description
G-201	WHITE FLAT, 1⅛″ O.D.
G-202	BLACK, FLAT, 1⅛″ O.D.
G-210	WHITE, BEV., 1$\frac{3}{16}$″ O.D.

Only with definite information as to model number, can your plumbing supply house substitute an accurate replacement, if he can't replace the manufacturer's original parts.

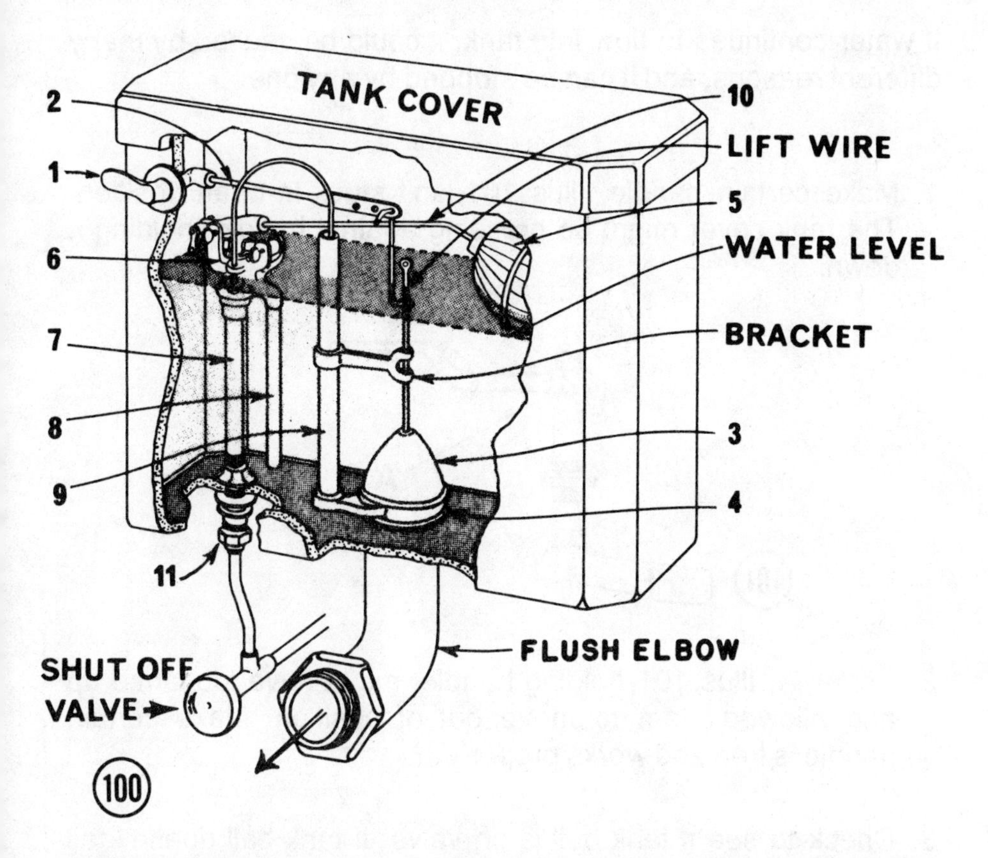

A toilet operates in the following manner. When you press handle 1, Illus. 100 it raises lever 2, which in turn lifts Tank Ball 3. Water in tank rushes down open valve 4, pours into bowl into waste line, either to septic tank or city sewerage system.

TANK ON TWO PIECE TOILETS

When water in tank flushes down toilet, the float 5 drops. This releases pressure on ballcock 6. Water, coming into tank through riser 7 flows through ballcock into hush tube 8. When tank ball closes valve 4, water raises float. When float reaches its specified height, it applies pressure and closes ballcock.

Everytime a new house is connected to the city water line, the water line picks up a lot of foreign matter. If dirt or sand gets in, it could lodge against ballcock washer 6, or on tank ball seat 4. This allows a trickle of water to flow into tank or bowl, Illus. 103.

If water continues to flow into tank, it could be caused by many different reasons, and it can be stopped by any one.

1. Make certain handle, Illus. 101, isn't stuck in open position. The tank cover might be pressing against handle, holding it down.

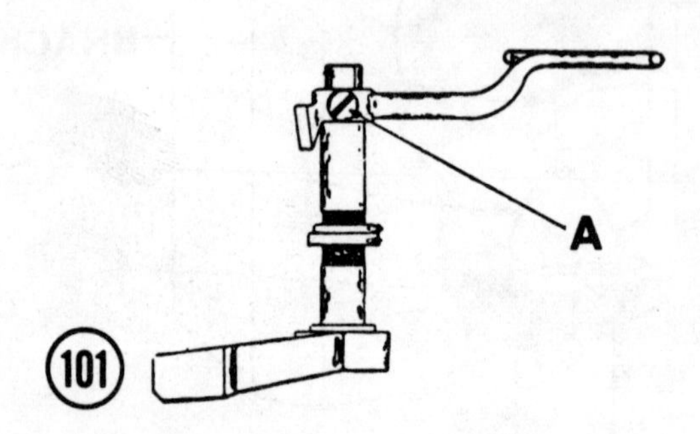

2. Screw A, Illus. 101, holding handle, might have loosened up and allowed stem to move out of position. Make certain handle is free and works properly.

3. Check to see if tank ball is on valve. If tank ball doesn't fall automatically into position on valve, it could be caused by rust on lift wire or in bracket. Remove lift wire and steel wool shaft clean. Apply vaseline to lubricate holes through bracket.

4. If tank ball is on valve, and water continues to flow, shut water off, flush toilet. Remove tank ball by unscrewing lift wire. If tank ball is worn, shows ridges, or rust buildup, replace ball.

5. Inspect flush valve seat 4, Illus. 102. Use a rag to polish rim. Remove rust, dirt, etc. A particle of sand on edge can keep tank ball from seating properly. After installing new tank ball turn water on to see if it shuts off automatically at proper water level. If water continues to run into overflow 9, Illus. 100, check to see whether float 5 is operable.

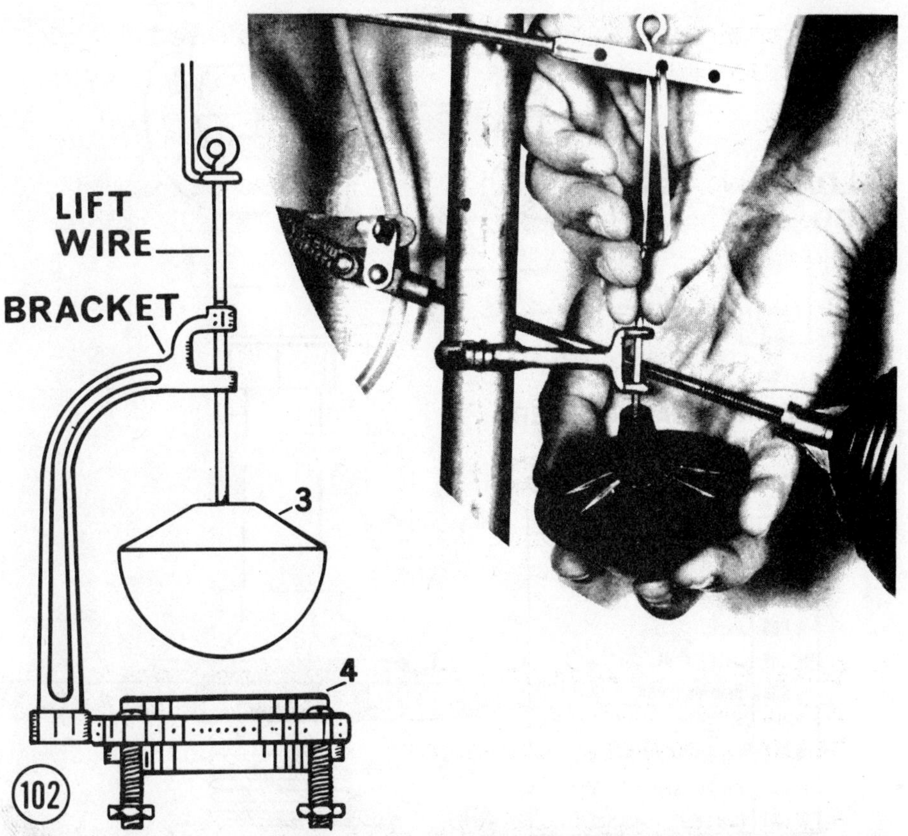

6. If float 5, is operable. Again shut water off, flush toilet, unscrew float. Shake float to see if it contains any water. If you hear any water in float, replace float.
7. If water continues to run, bend float arm 10 down slightly. This should automatically shut water off at a level lower than that indicated on inside of tank.
8. If water continues to run, again shut off water, flush toilet and remove ballcock, Illus.103. This is done by loosening thumb screws 1 and 2. Slide lever out, remove ballcock, Illus. 104.

To replace ballcock assembly, Illus. 103, disconnect coupling nut #11, Illus. 100. Remove float and rod #10. Disconnect lock nut #4, Illus. 103; lift out assembly. Install replacement making certain it has the same length shank #5, Illus. 103. Using pliers, Illus. 10, tighten lock nut only enough to compress and make shank washer #3, Illus 103 watertight. Use extreme care as too much pressure could crack fixture.

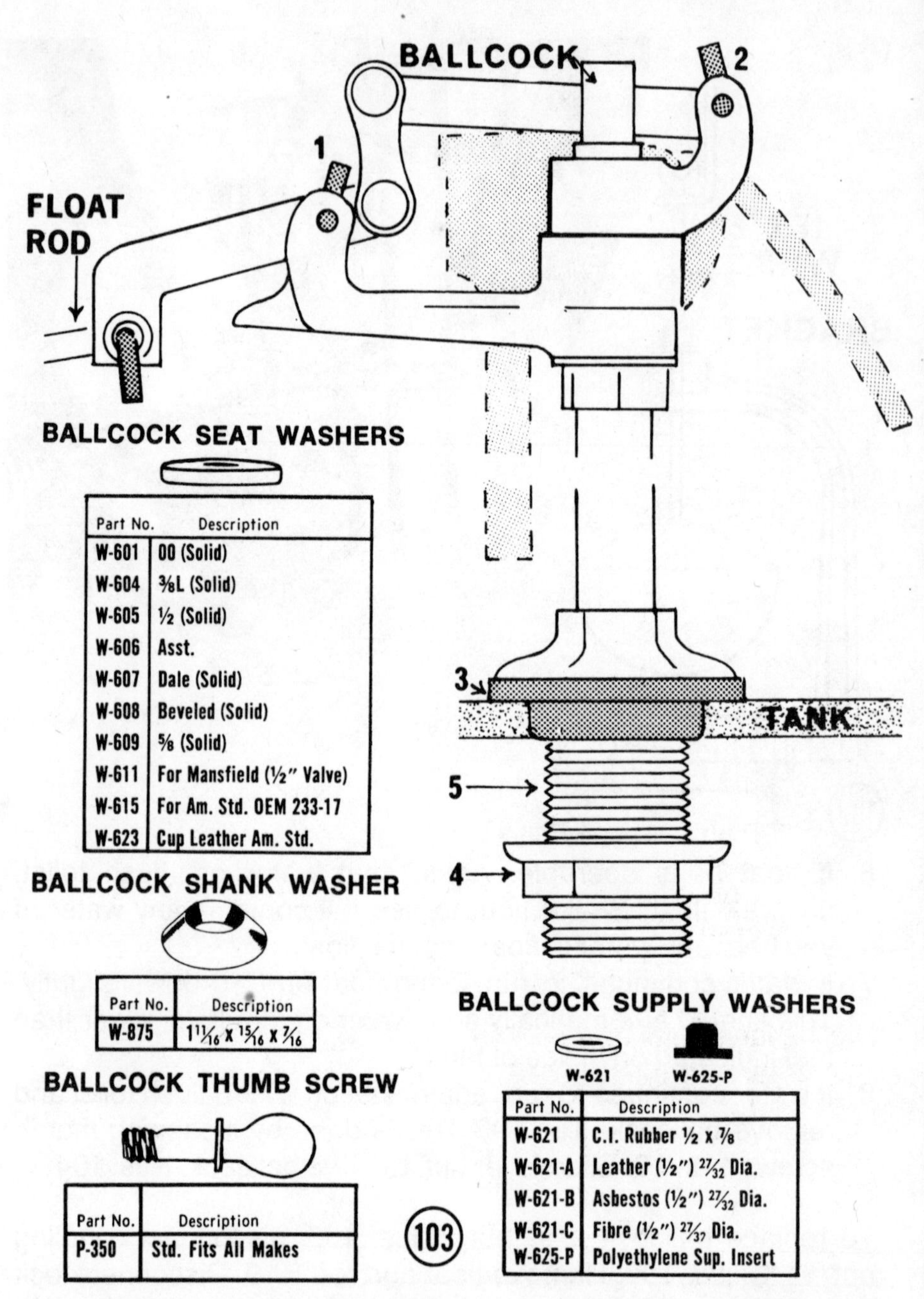

BALLCOCK SEAT WASHERS

Part No.	Description
W-601	00 (Solid)
W-604	⅜L (Solid)
W-605	½ (Solid)
W-606	Asst.
W-607	Dale (Solid)
W-608	Beveled (Solid)
W-609	⅝ (Solid)
W-611	For Mansfield (½" Valve)
W-615	For Am. Std. OEM 233-17
W-623	Cup Leather Am. Std.

BALLCOCK SHANK WASHER

Part No.	Description
W-875	1 11/16 x 15/16 x 7/16

BALLCOCK THUMB SCREW

Part No.	Description
P-350	Std. Fits All Makes

BALLCOCK SUPPLY WASHERS

W-621 W-625-P

Part No.	Description
W-621	C.I. Rubber ½ x ⅞
W-621-A	Leather (½") 27/32 Dia.
W-621-B	Asbestos (½") 27/32 Dia.
W-621-C	Fibre (½") 27/37 Dia.
W-625-P	Polyethylene Sup. Insert

Since there are many different ballcocks in service, yours may differ. It should still have most of the same working parts. When you know "which does what," you make necessary adjustments. A good investment, even before you need it, is a ballcock repair kit that fits your ballcock.

To replace a washer on a ballcock, Illus.104,105,loosen screw at bottom and replace washer with exact size.

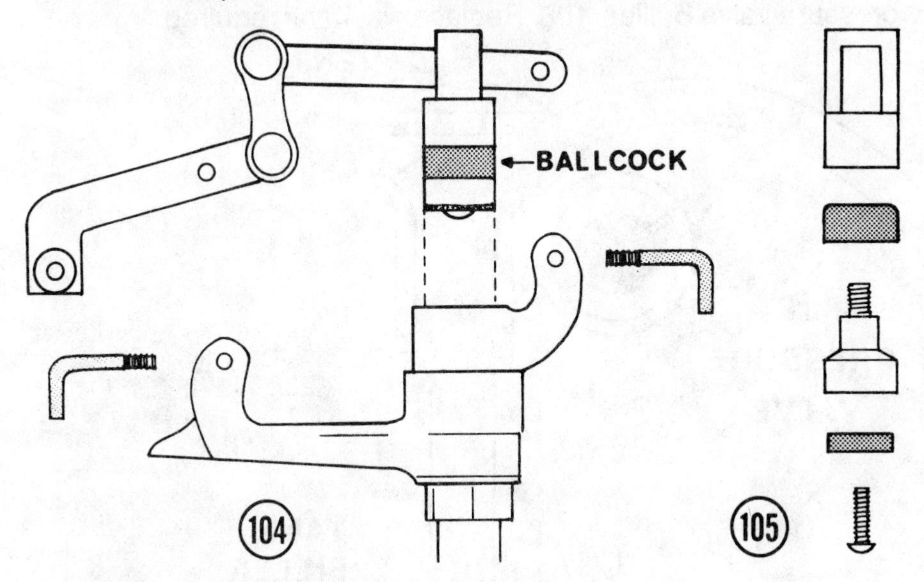

If ballcock has a ring nut on bottom, Illus.106,use care not to burr nut when removing. To protect, wrap edge with tape. Take worn washer to store to make certain you obtain exact kind and size. If it's not possible to obtain washer required, turn worn washer over, place good face down, but by all means, get a replacement as soon as possible.

Replace split washers, Illus.107with same kind. Look down into seat of ballcock to see if there's any rust, sand or particles of dirt cemented to seat. Clean thoroughly, apply vaseline and replace. Again test toilet.

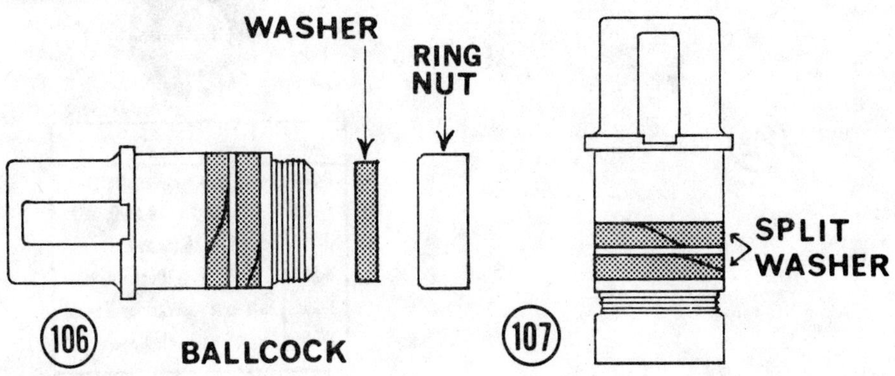

If water coming into tank creates loud noises, you can frequently eliminate noise by decreasing pressure. To do this adjust pressure valve B, Illus. 108. Replace washer if required.

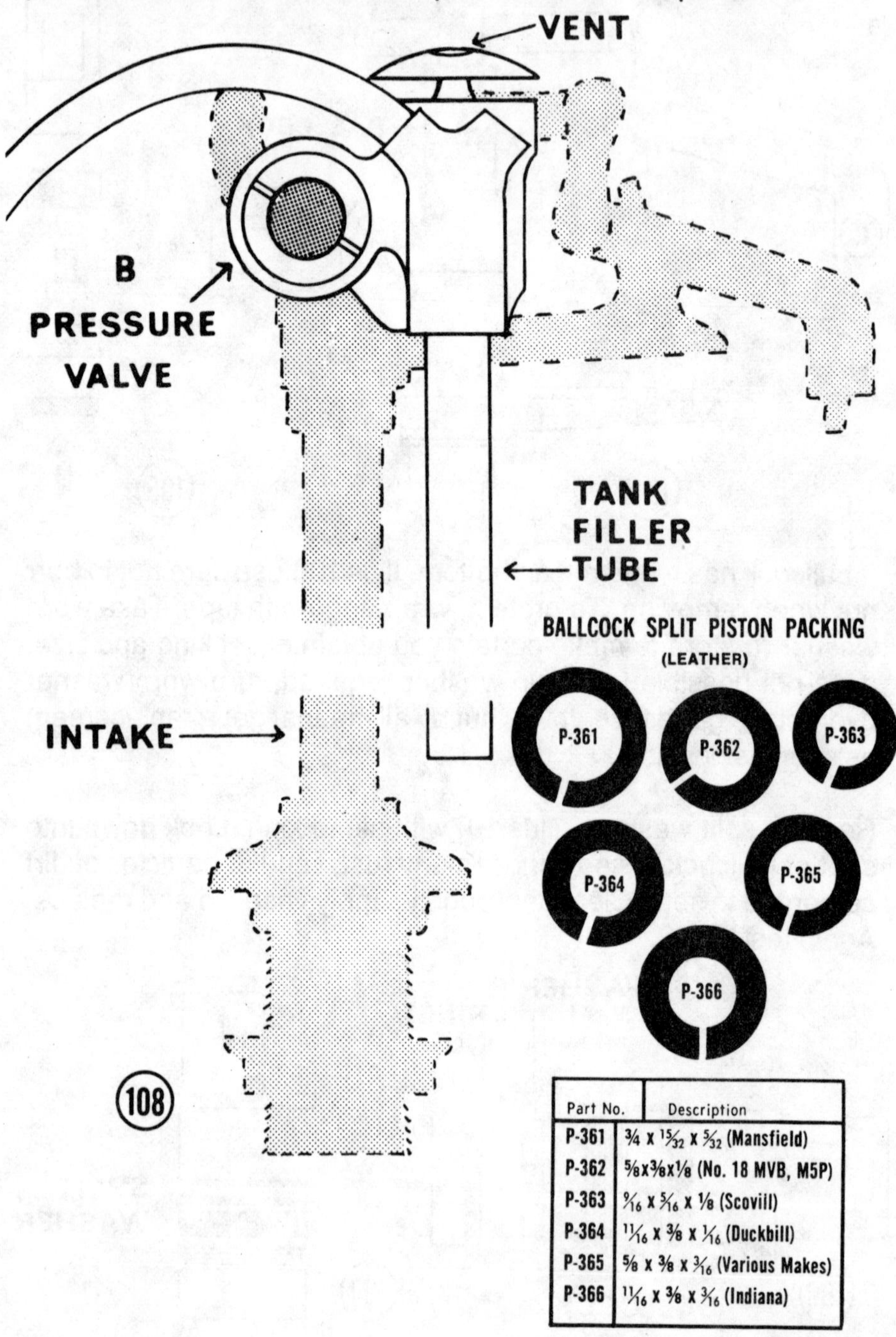

Part No.	Description
P-361	3/4 x 15/32 x 5/32 (Mansfield)
P-362	5/8x3/8x1/8 (No. 18 MVB, M5P)
P-363	9/16 x 5/16 x 1/8 (Scovill)
P-364	11/16 x 3/8 x 1/16 (Duckbill)
P-365	5/8 x 3/8 x 3/16 (Various Makes)
P-366	11/16 x 3/8 x 3/16 (Indiana)

If water spouts out of vent, Illus.109, shut water off, flush toilet. Loosen screw at top. Remove cap and unscrew vent A. Clean rust, replace washer. Also clean rust in chamber. Apply vaseline and replace.

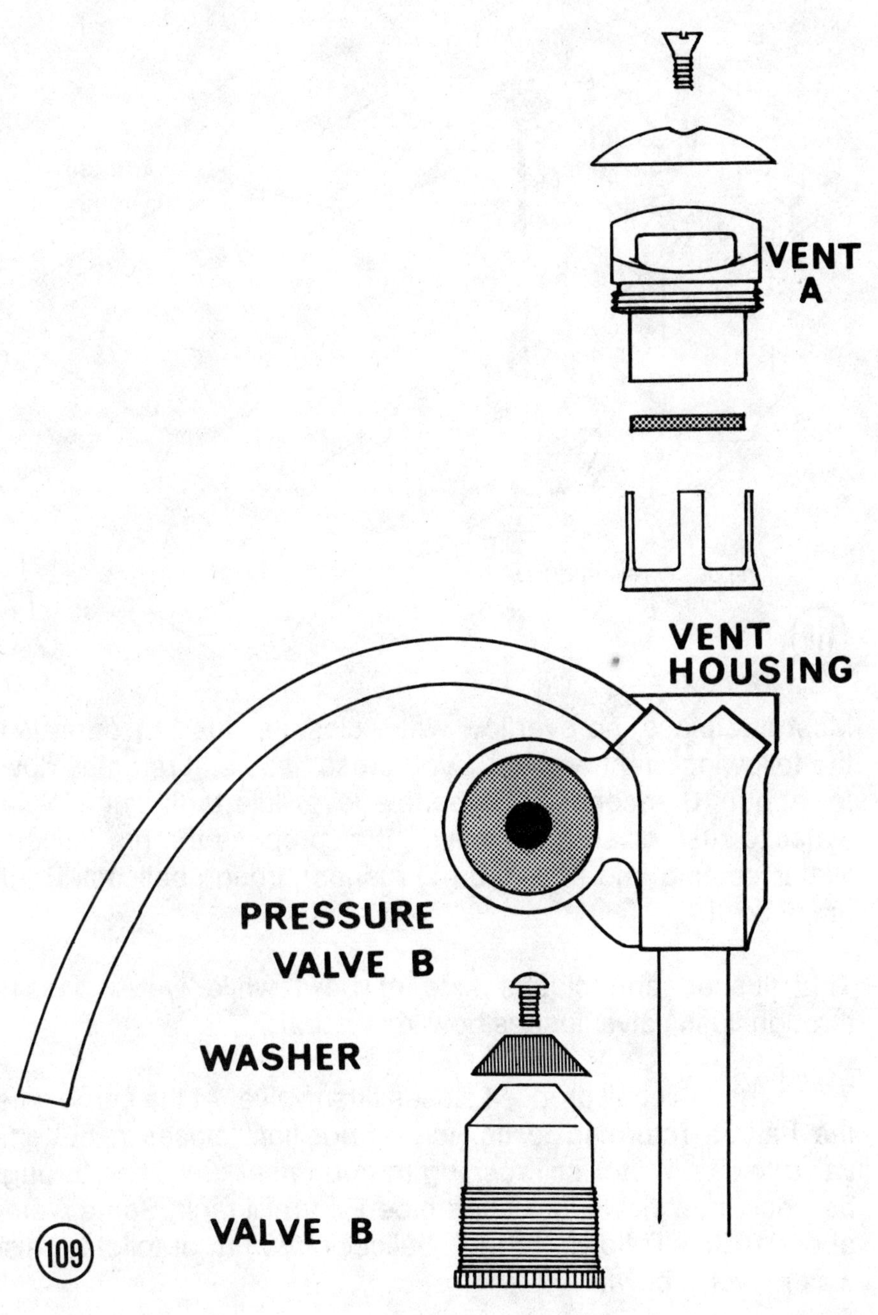

Illus.110 shows a rim type pressure valve. Screw A, holds ring in position. Loosen A and turn knurled rim B clockwise to cut down pressure.

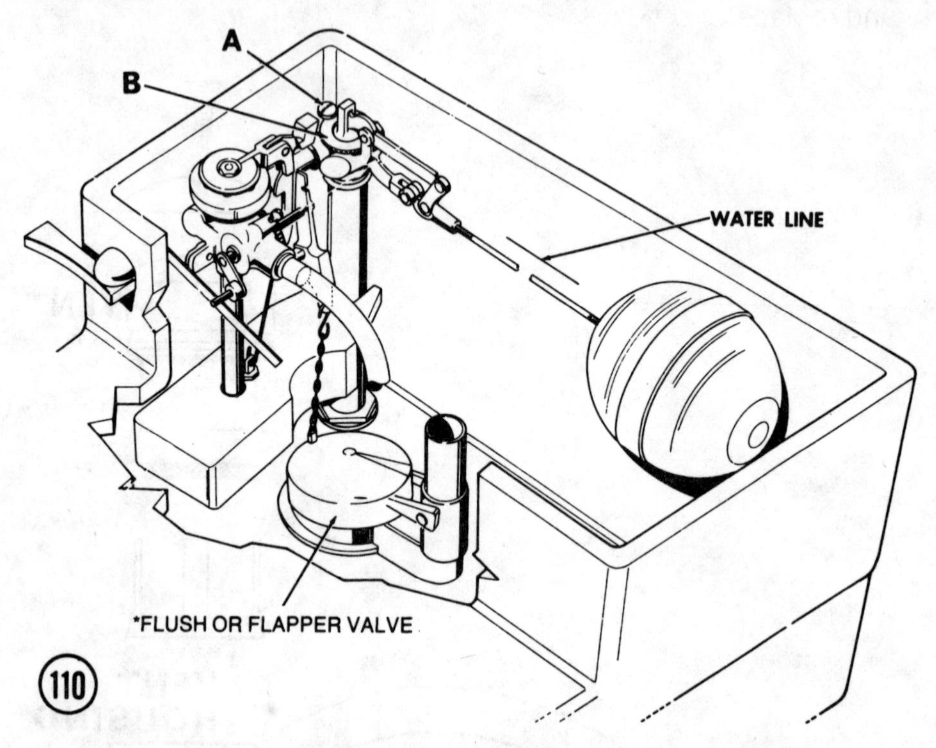

Most one piece, no-overflow water closets, Illus.111, operate in the following manner. When you press handle, it remains down for about 10 seconds. The handle lever lifts tank ball allowing water to rush down flush valve*; it also opens the rim valve C. Water coming into riser tube D rushes through ballcock E into rim of bowl.

This flushes and cleans side of bowl while water passing through flush valve flushes bowl down drain.

When the tank ball drops, it closes flush valve. At the same time, the handle returning to its normal position, closes a butterfly valve to rim. Water still rushing through riser tube, and through ballcock, now flows into hush tube F to refill tank. Some water, about 10%, still flows through ballcock into rim of toilet to raise water level in bowl.

When water raises float to proper level, it closes ballcock. The hard rubber ball in vent G, Illus. 111, drops into position and prevents water from siphoning from tank into bowl.

If incoming water spurts out of Vent Cap and leaks out of tank cover, adjust water pressure at B. Also inspect and replace rubber washer in Vent Cap, replace Vent Ball. If you can't get a replacement ball and washer, put an empty aluminum foil cheese container over cap in position shown. While it should stay in place, a dab of contact cement on inside of tank cover, or on bottom of container, should keep it in exact position.

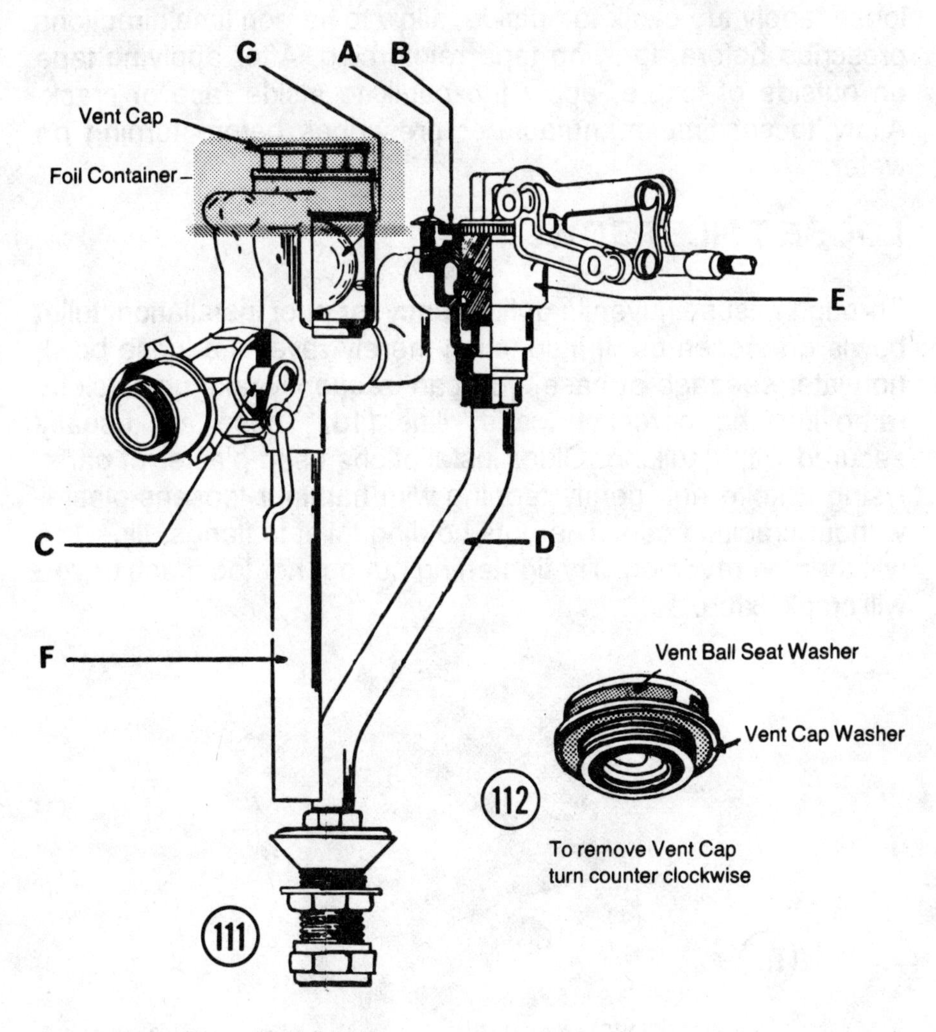

CRACKED TOILET TANK OR BOWL

If a toilet tank is damaged, and the break leaks water, you can frequently make a temporary repair using a silicone tub caulk now on sale in most hardware stores. Shut off water to fixture. Drain and sponge dry. Thoroughly clean area around crack before applying sealant. Apply sealant exactly as directions specify. Be sure fixture is in a warm room. Don't attempt to repair a cold fixture with a sealant that requires 70° to set up.

If fixture has a through break, or large particles have come loose, apply tub caulk to outside, allow to harden time directions prescribe before applying tape reinforcing. After applying tape on outside of fixture, apply tub caulk to inside face of crack. Allow to set time manufacturer prescribes before turning on water.

LOOSE TOILET BOWL

Through misuse, juvenile delinquency, or poor installation, toilet bowls do loosen up. If inspection merely reveals a loose bowl, no water seepage at base, you can frequently take up slack by removing the porcelain caps, Illus. 113. These are usually secured with caulking. Older installations used plaster of paris. Using a knife and gently tapping with hammer loosens plaster without cracking cap. The nuts holding toilet to flange, Illus. 114, will then be revealed. Try tightening nut but not too much or you will crack fixture.

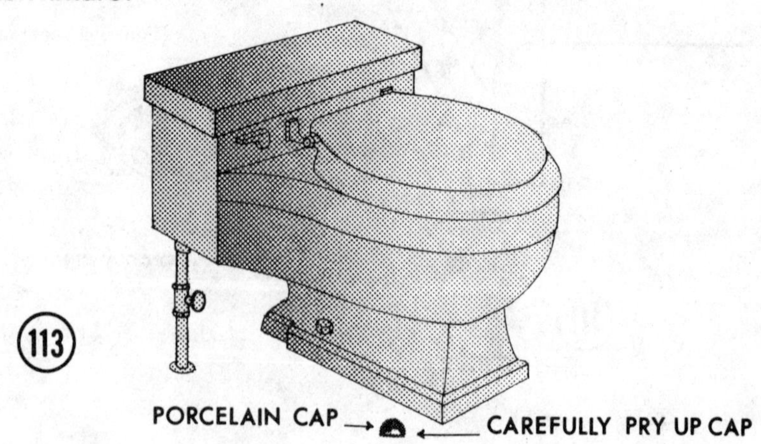

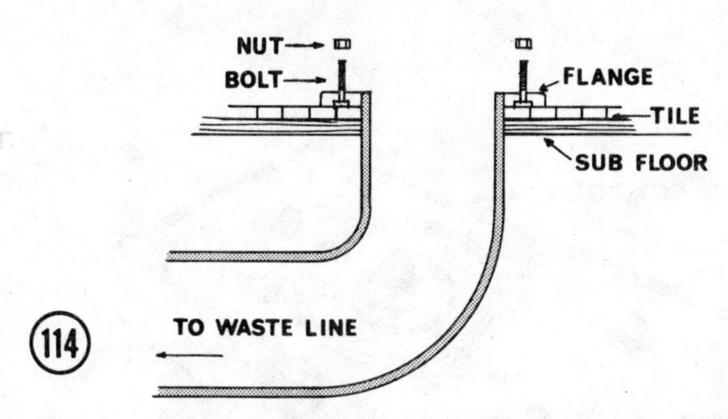

If water is leaking at base of a one piece toilet, shut water off, flush toilet, sponge out water, disconnect water line to tank. Loosen and remove nuts, lift toilet straight up. Scrape away old seal. Buy a new seal, Illus. 115, from your plumbing supply house.

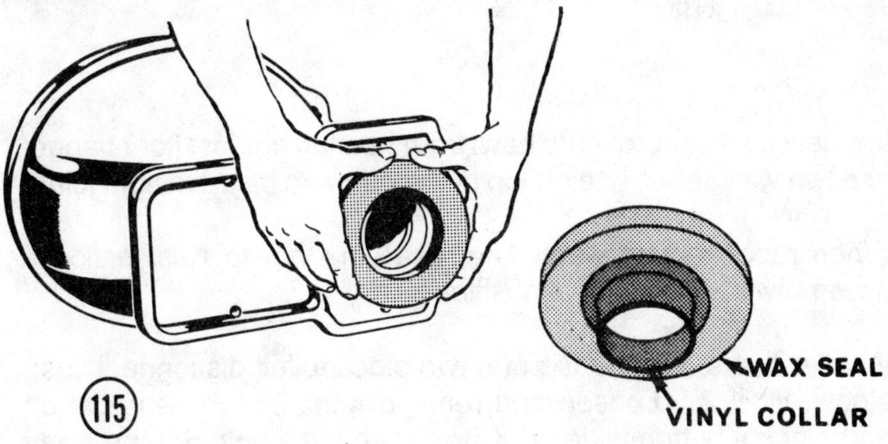

HOW TO REPLACE TOILET BOWL SEAL

To replace a seal, Illus. 115, apply setting compound, available from your plumbing supply house, Illus. 116. Place new seal in position on horn of toilet and lower toilet directly over flange on floor. Don't come in at an angle. Don't slide it into place. Come straight down. Press toilet down with a slight twisting motion. Be sure to aim toilet so bolts in flange come through holes in base. Fasten nuts snug, but do not tighten with force. Apply setting compound around nut and replace caps. Reconnect water line using Tape Dope on male threads.

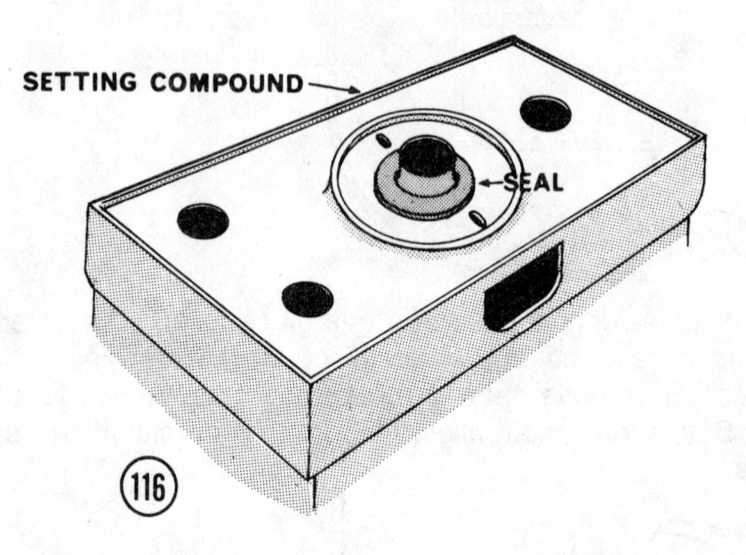

If a new tile floor prevents bowl from seating against floor flange, use two wax seals, one on top of the other, to insure a tight joint.

When reconnecting water line, use care not to twist ballcock assembly from its original position.

If water leaks out at base of a two piece toilet, disconnect flush elbow, Illus. 37. Loosen and remove nuts, lift toilet straight up and place it bottom side up. Carefully handle unit. Scrape away setting compound on edge of toilet, also remove old seal.

Place a level on flange and on floor where toilet sets. If floor has settled, it may be necessary to shim toilet. Use small pieces of wood shingle placed close to bolts. Replace toilet and check with level. If it checks OK, carefully lift toilet out of position without disturbing shims. Lay a thin bead of plaster of paris on rim. Use a thicker bead alongside a shim. Replace toilet seal. Press toilet into seal and check with level. Only tighten bolts snug. Use extreme care not to crack fixture when tightening bolts. Remove excess plaster of paris. Clean edge of rim with a damp rag.

LOOSE TOILET SEAT

This can usually be traced to a worn rubber washer, Illus. 117, or worn rubber bumpers, Illus. 118.

Remove tank cover, shut off water, flush toilet. Use end wrench to loosen nuts. Remove seat and bolts. Replace washers on shaft with beveled face facing tank. Insert bolts in holes, apply washers and nuts. Tighten snug but do not force.

CLOSET SEAT BUMPERS

These are made with screws or nails. If nail type, use pliers, pull out worn bumpers, replace with new ones, but not in the same holes. Use same size.

TACK BUMPERS

Part No.	Description
B-306	½" WHITE
B-307	⅝" "
B-308	¾" "
B-314	½" BLACK
B-315	⅝" "
B-317	¾" "

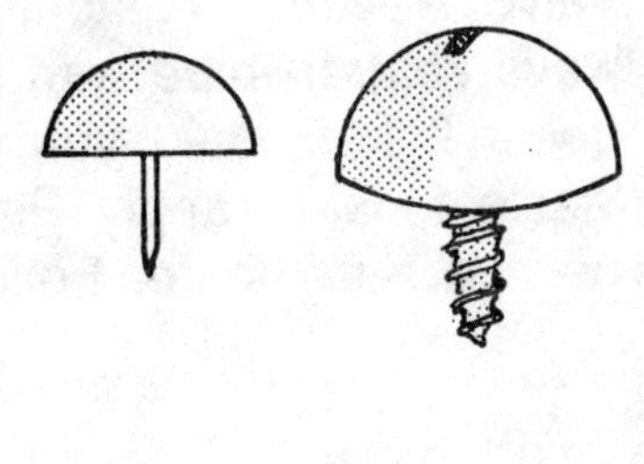

BAR BUMPERS

Part No.	Description
B-101	1⅝" BLACK
B-102	2¼" "
B-105	2¼" WHITE
B-107	1½" "
B-202	2⅛" WHITE (TACK)

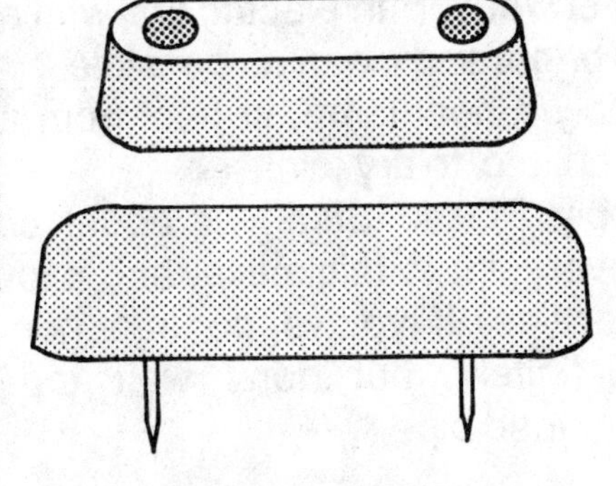

REPLACE TOILET SEAT

Measure spacing between seat bolts and buy replacement that fastens in same position. If present seat bolts go through tank, be sure replacement bolts fits this installation.

It isn't necessary to shut off water, or flush toilet, but since there's a good chance of dropping a nut, washer or wrench into tank, shut water off, and flush toilet.

If nuts holding bolts have rusted, use liquid wrench to loosen. Allow oil to penetrate. Tap nut lightly and it now should turn freely. Don't tighten nuts with force. You will crush rubber with too much pressure, thus shorten its life span.

LEAKY PIPE JOINT

If a leak occurs where a valve or faucet is connected to pipe, first try to tighten joint with wrench. Use a stillson to hold pipe, an end wrench on valve. Never try tightening a valve using only an end wrench. Always support pipe with a stillson or pipe wrench. If tightening doesn't stop leak, you can try three other repairs. First, and easiest, is to apply Plastic Steel or equal quality compound with a putty knife. Follow this step-by-step procedure.

1. Shut water off to joint.
2. Relieve pressure in joint by opening a faucet. Drain line by also opening a faucet downstairs.
3. Clean rust, scale, dirt or other foreign matter off joint. Use steel wool or an electric drill with a wire brush. Polish the joint as bright and clean as possible.
4. Apply a heat lamp, or heat from a large light bulb, sun lamp, torch, etc. to dry joint.
5. Apply Plastic Steel, or equal compound with putty knife. Cover joint to thickness, and allow to set time manufacturer specifies. If you've done a good job, the repair will last indefinitely, but don't wait too long before replacing a damaged pipe.

If tightening a threaded pipe doesn't cure leak, and surface sealant doesn't work, take the joint apart. Clean off old compound on both male and female and apply a strip of new all-purpose Tape Dope® over male threads, Illus. 119. Tape Dope comes in a roll, like electric tape. Wrap Tape Dope once around, plus 1/2" on male thread. Reassemble joint. This makes a quick, easy, tight joint. Tape Dope is available at your plumbing supply dealer.

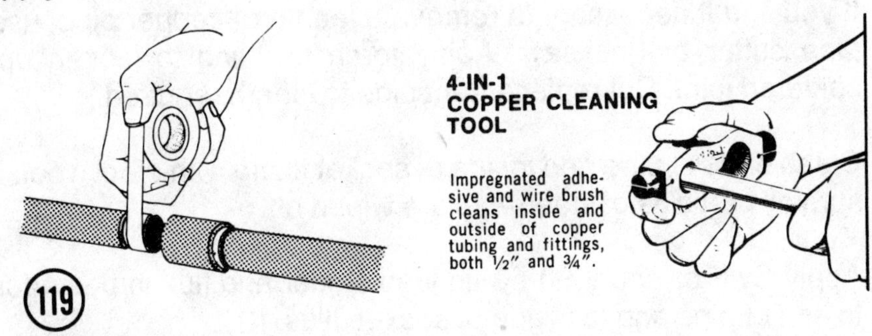

If pipe is 2" or larger in diameter, wrap Tape Dope at least twice around.

You can also repair a fine leak between a fitting and pipe, or in a cracked pipe or fitting, by soldering. Dry and clean a crack thoroughly as previously described. Apply Swif,® or equal solder, Illus. 120. Apply heat adjacent to crack. Do not apply flame on Swif solder. Use wire solder and flux to fill a large crack. Swif solder is easy to use. It contains flux. Brush Swif solder on following manufacturers directions.

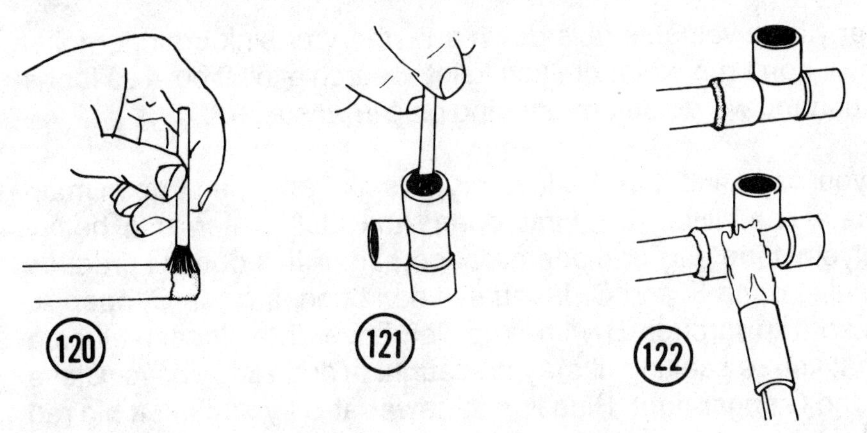

Solder penetration of one-third the cup depth — breaking load, approximately 2100 lbs.

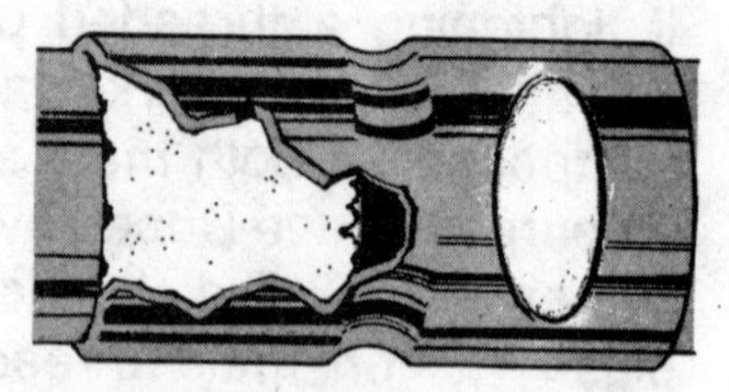

Solder penetration of the entire cup depth—breaking load, approximately 7000 lbs.

(122)

If you find it necessary to remove a length of copper pipe, use a pipe cutter or hacksaw. Apply torch to fitting to loosen up a soldered joint. Cut replacement pipe to length required.

Clean ends of pipe and inside of socket fitting with steel wool. Be sure all particles of steel wool are wiped off.

Apply Swif or equal 50-50 tin lead solder and flux in paste form to end of pipe and to inside of socket, Illus. 121.

Join pipe and fitting, Illus. 122. Allow the small "collar" of solder paste to remain as shown. Apply torch to pipe and fitting but not directly on solder. When fitting and pipe heat up the solder turns from metal grey in color to black. It will now start bubbling. Remove flame and allow to cool. The heat draws the solder into joint. When bubbling stops, use a damp cloth to brush joint clean.

LOST & FOUND DEPARTMENT

If anything valuable falls down a lavatory or sink drain, or toilet bowl, don't run water or flush toilet. Search a toilet bowl as far as your hand will go before opening plug in waste line.

If you don't find it in the lavatory or sink trap, open trap in main waste line, Illus. 8 . First open A. Catch water in a heavy polyethylene bag or clean garbage can. If this doesn't produce results, open B and C. If you still don't find it, close C, open A. Flush trap through B with a garden hose. If this doesn't locate what you're seeking, it may be caught in drain line above. Close A and C finger tight. Plug line to sewer at B by stuffing a big rag in opening. Half fill bath tub with water and on signal, have someone open tub drain and flush a toilet. With waste line full, open plug A, water rushing through drain and trap should dislodge any small object while it fills several barrels with water in your basement. If this doesn't locate the missing object, file a claim with your insurance agent.

FROST FREE OUTSIDE SILLCOCK

If winter arrives before you have shut off an outside faucet, a frost free, self-draining sillcock, Illus. 123 , will eliminate a freeze-up. Always remember to disconnect and drain a garden hose.

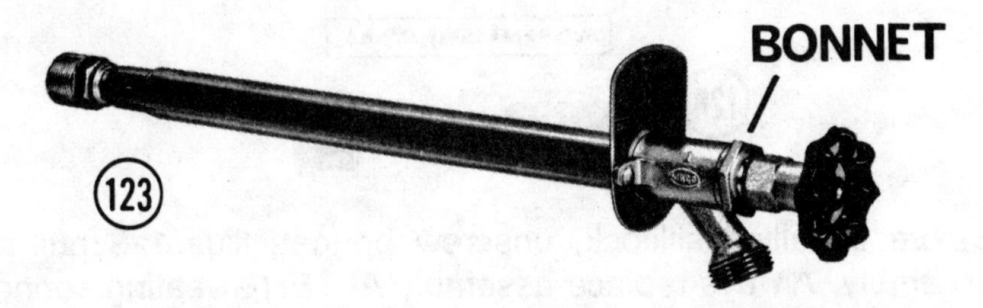

Select sillcock length required, Illus. 124, to project through wall, so valve seat, Illus. 125, is within a heated area. Sillcocks come in 6", 8", 10" and 12" lengths. Drill a 1" hole through foundation. The sillcock shown in Illus. 123, is available with a Universal connection, Illus. 126 . Threads A take a 3/4" M.S.P.S. Threads B take a 1/2" F.S.P.S. If you remove the Universal adapter A., you can sweat an adapter, Illus. 127 .

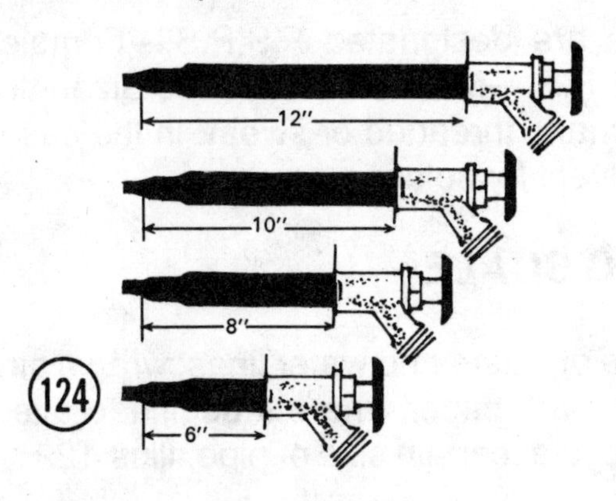

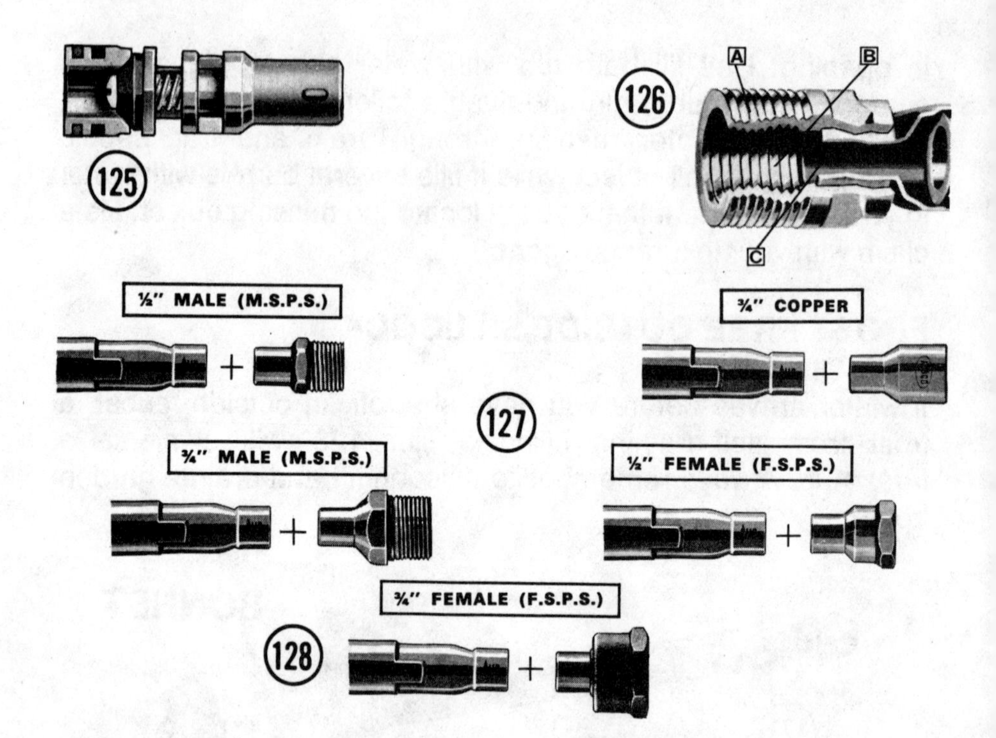

Before installing sillcock, unscrew bonnet, Illus. 123, pull out assembly. Always replace assembly AFTER sweating connection to pipe.

If you want to connect copper tube to equipment having standard iron pipe connections, copper fittings are available with either male or female threaded ends, Illus. 128.

FACTS ABOUT FITTINGS

Threaded fittings are designated F.S.P.S.—Female Standard Pipe Size; or M.S.P.S.—Male Standard Pipe Size. All fittings are either male or female, threaded or sweat. In the case of plastic, you substitute solvent for solder.

HANDY SIZING SCALE

To determine size of waste and water lines, wrap a strip of paper around pipe and mark paper where it begins to overlap. Place strip against scale to ascertain size of pipe, Illus. 129.

HANDY PIPE SIZING SCALE

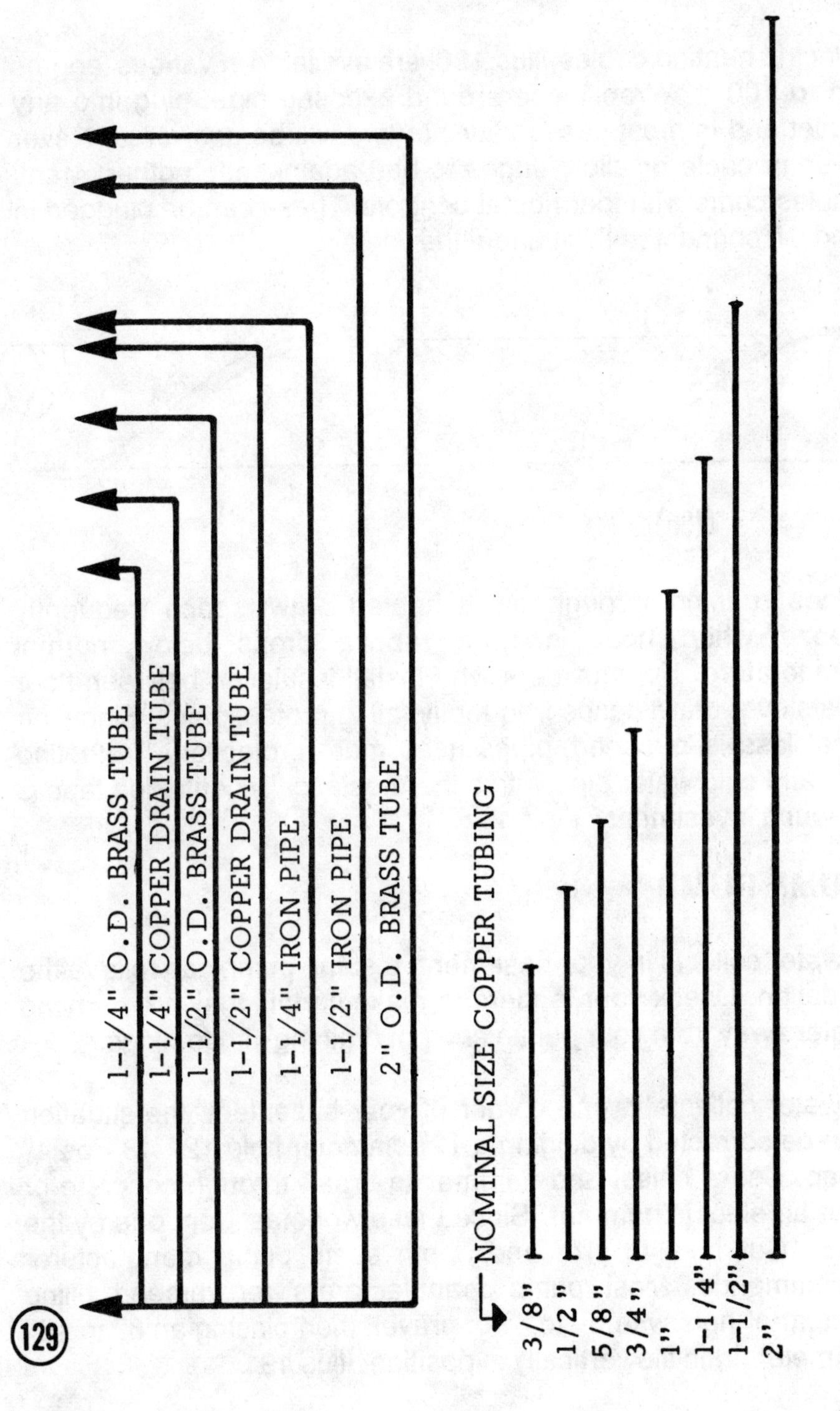

THAWING FROZEN PIPES

Electric heating cables, Illus.130, are available in various lengths up to 100 ft. Wrap these around exposed pipe, plug into any outlet and in most cases your prayers will be answered. Never overlap cable or allow edges to butt against each other. Many cables come with thermostat controls. These can be plugged in and left connected throughout the winter.

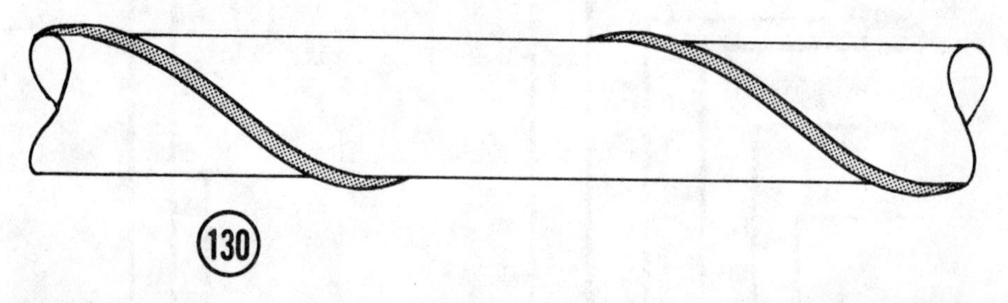

(130)

Pipes running through an unheated crawl space frequently freeze when heat in room above drops below normal temperature. Homeowners who install insulation between floor joists over crawl space frequently fail to protect pipes. Since the heat loss is lessened, pipes need more protection. Protecting hot and cold water pipe with a thermostatically controlled tape is a sound investment.

SUMP PUMP

If water collects in your basement, a sump pump can relieve the situation. Check your leaders to make certain they are carrying water away from your house and not draining into footings.

If water collects in one corner of your basement, the situation can be corrected by digging a 12" diameter hole 12", 18" to 24" deep. Use a chisel and hammer to break through concrete or rent an electric hammer. Since these work fast, rent one by the hour. Cut hole to size and depth sump pump manufacturer recommends. Most pump manufacturers recommend filling bottom of hole with 3" to 4" of gravel, then placing an 8" to 12" diameter drain tile vertically in position, Illus.131.

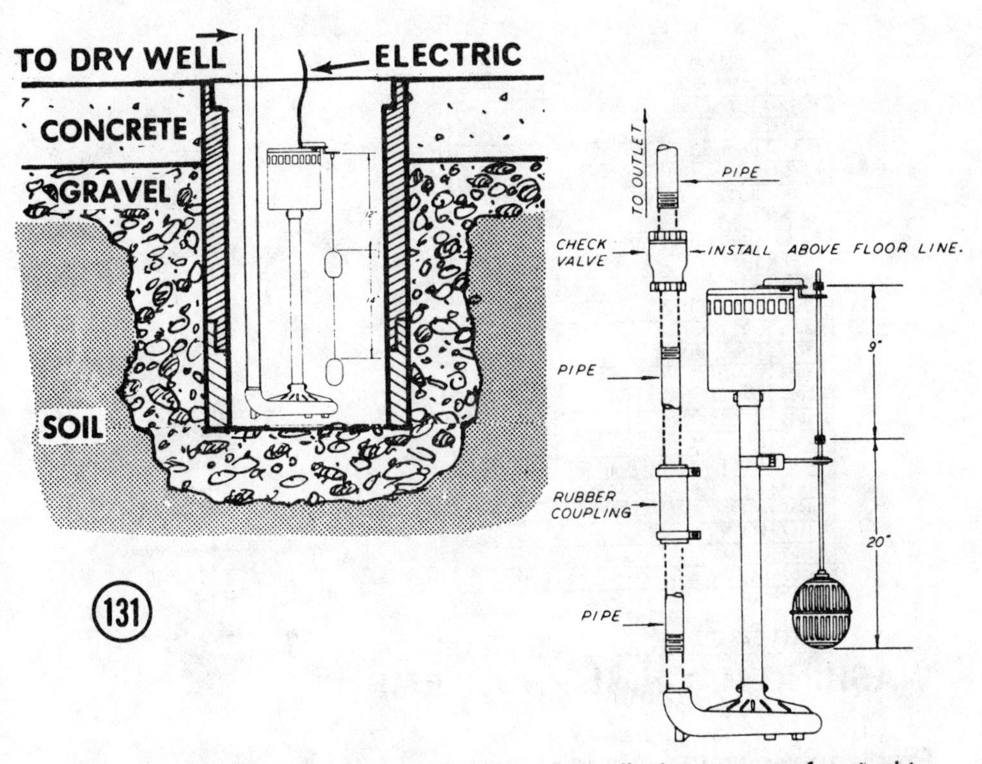

Drain tiles designed for sump pump installation are perforated to allow for seepage. Back fill around tile with 3" to 4" of 1" to 1-1/2" crushed stone. If you can't buy a cap for the tile, make one from 3/4" exterior grade plywood. Cut and glue two thicknesses to make a 1-1/2" thick cap. Cut discs to inside diameter of tile, A, Illus.131. Top of tile should finish flush when you repair floor.

Drill holes in cap, one for pipe, another for electric conduit, plus several more to allow circulation of air to pump. Connect water line from sump pump to a dry well. Since you will have to go through foundation wall to discharge water, drill hole through foundation at highest possible point. You can use plastic pipe. When buying pump, be sure to find out whether it will lift water to height your basement wall requires.

If your basement is damp, install a humidifier. Select a size with sufficient capacity to handle overall size cubage of your basement.

PART NO.	DESCRIPTION
P3041	DISPLACEMENT WEIGHT ASSEMBLY
P3324	SET SCREW, COUPLING
P3374	BUSHING
P3616	IMPELLER
P3623	COUPLING, MOTOR
P3781	BOTTOM PLATE, HOUSING
3798	TUBE, RISER
P3799	SCREW, BOTTOM PLATE
3802	SHAFT
P3830	ROD, FLOAT
3832	ROD, FLOAT
P3835	BOTTOM PLATE, HOUSING
3836	TUBE, RISER
3856	HOUSING, CAST IRON
3857	HOUSING, BRASS
P4010	SET SCREW
P4182	BALL, FLOAT
P4483	TOP PLATE, HOUSING
P4484	TOP PLATE, HOUSING
P4485	SCREW, TOP PLATE
P4557	STRAP, CLAMP
P6002	MOTOR
P6003	MOTOR
P6030	GUIDE, ROD
P6031	SET COLLAR
P6148	SCREW

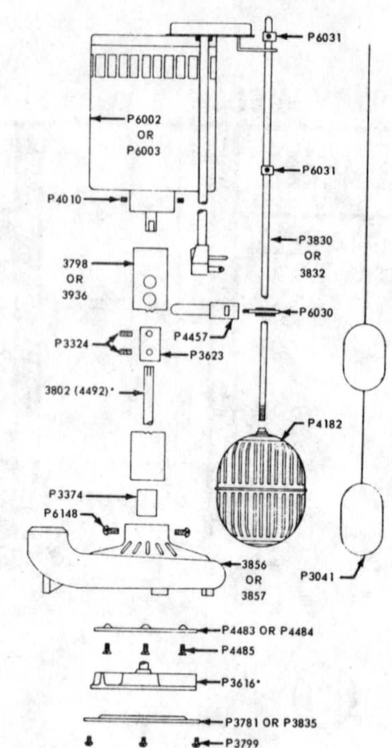

WASHING MACHINE PROBLEMS

Problem: No water entering machine.
Solution: Check to see if someone shut off valve to washing machine line.
Check to see if water pressure is OK at other faucets. If OK, the probable fault could be:
1. A clogged strainer, Illus. 132.
2. A loose washer on end of faucet spindle, Illus. 26.

When new homes are being erected in a neighborhood, mud and other foreign sediment frequently builds up. Remove hose from faucet and remove strainer. If clogged, replace. Your hardware store has replacements. If strainer is OK, or replaced, and the problem still exists, check washer on end of spindle.

132

WATER TANK REPAIRS

While pin holes in a galvanized water tank indicate its life span is nearing an end, you can make a quick repair by screwing in a self-tapping plug, Illus.133. These come in various sizes. Always buy at least three different sizes. Use the smallest size first. Using an adjustable end wrench, just screw it in. If the spot is weak around the edges, the plug won't seat itself. Use the next size plug until you seat a plug securely in position. Keep these plugs handy as you can make an instant repair. This will give you ample time to shop for a replacement tank.

An ace plug, Illus.134, simplifies repairing a larger hole.

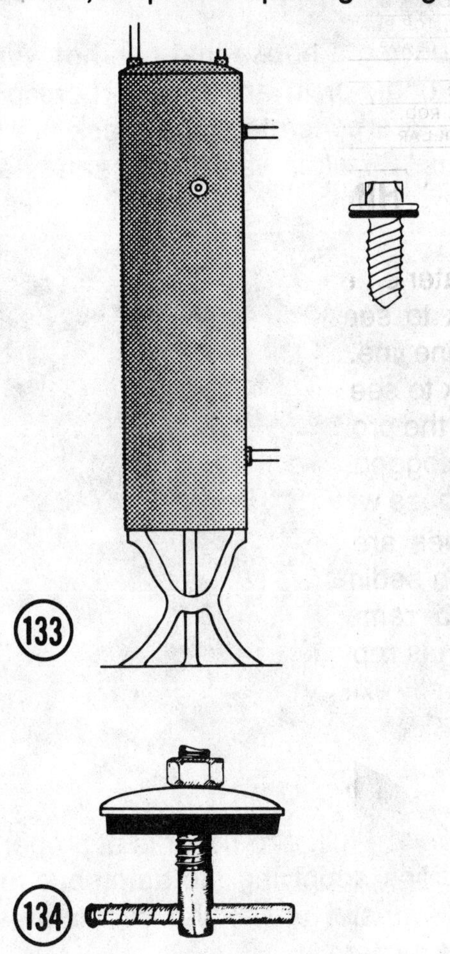

BOILER SEAL *Powder*

Seals leaks in boilers and heating systems, in steam and hot water systems, pipes and radiators. Will withstand the same heat and pressure of the boiler itself.

BOILER SEAL *Liquid*

Companion to Powder Boiler Seal to be used whenever a liquid seal is required. Will not clog system ...recommended for all new installations.

HOT WATER HEATING PROBLEMS

If you consider buying a house that has hot water heating in a downstairs playroom, or in any room where pipes may have been embedded in a concrete slab, check the water pressure boiler gauge, Illus.135, after water has been heated to required temperature.

(135)

When all radiators are full, and boiler is at proper pressure, shut the valve on the line supplying the automatic feed valve, Illus. 136. Turn the thermostat up so the circulator starts pumping water through the system.

136

A hot water pipe, buried in concrete, can spring a leak and still permit the system to operate satisfactorily. While an automatic intake valve will normally keep the pressure up, the burner will run longer, and at more frequent intervals. A leaking system can be operative with no one knowing a leak exists.

Since most hot water heating supply and return lines, Illus.137, bedded in a concrete slab, are usually laid over a filled-in area, sometimes covered by #15 felt, sub and finished flooring, water frequently sinks without appearing on the surface. Close the intake supply line and the boiler gauge will drop. Unless you have a map showing exactly where the supply and return lines are buried, a repair can be a costly and frustrating job.

137

RUST-RAIDER

Rust-raider turns boiler water blue. If water changes to any other color, add Rust-raider until water turns blue.

The hot water line from boiler is connected to a circulator. The circulator supplies radiators through a main supply line plus feeder lines, Illus. 138, or from another radiator. Since the water is constantly being pumped through the system, then back to the boiler through a return line, all radiators maintain constant heat, except when an air block occurs.

Many baseboard radiators have automatic air bleeders while the older ones, Illus. 139, require loosening valve A with a screwdriver. Air will hiss out of B. Since water will also spurt out, keep a cup or glass in position to catch it. When you remove air, and only water spurts out, the radiator is full.

When a leak occurs in either the supply or return lines, or in any feeder, air will continually enter the system and you'll hear a gurgling or knocking.

138

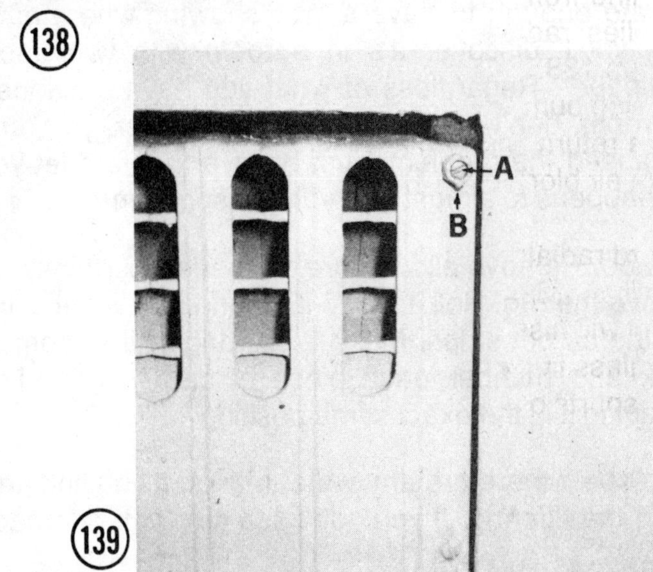

139

When you have located the leak, switch burner off, close intake supply valve, open bleeder valves A,Illus. 139 , on all radiators. Drain water from boiler. Drop pressure gauge to 0.

If you are purchasing a home that has hot water heating, make certain the seller supplies a map, or obtains one for you prior to purchase. Learn where and how deep hot water heating pipes are buried. Also find out where the main water and waste lines are buried. Some may be buried alongside, under or cross over heating lines.

Finding a buried pipe can be one of the most exasperating experiences a homeowner can go through, and anyone out to make a fast buck can legally rob you blind looking for a pipe.

Regardless of how much you value your time, no matter what you earn as a skilled craftsman or professional, the hourly rate charged for repairing a buried pipe will cost a sum far in excess to what you earn. If at all possible take a day or two of your vacation time and spend it on the job.

No matter what month you were born, and what luck you might have had in the past, whenever a buried pipe begins to leak, it's usually under your finest finished flooring or newest carpeting. If you are fortunate enough to have a map showing where each line is buried, you'll discover it's in a room with your most expensive wallpaper. Regardless of what you have previously spent on this room, don't lose your cool. Remember, you, and no one else caused the problem. Just believe that fate let you have one that happens to almost everyone at sometime.

Here's what to do. Remove all furniture, draperies, curtains, etc. Carefully remove the rug. Roll it, wrap it up and take it out of the room. Pry up the carpet strips. Make a drawing of the room so you can mark and number each piece of carpet strip. This simplifies replacing it in the exact same position.

Inspect floor to see if there are any water marks. If you find any, start tearing up the flooring. If you don't see any water damage do this.

As every homeowner who lives with a leaky roof soon learns, water flows in many mysterious and sneaky ways. When a pipe in a slab springs a leak, unless you are very rich, don't call in a plumber or heating contractor. Go first to your doctor and borrow a stethoscope. Ask him to show you how to use it, then probe your floor. Start probing where a hot water heating line enters the concrete. With the thermostat set high, and the circulator pumping water, probe the floor in the area where you think the pipe is buried. When you hear gurgling, you begin to zero in on the leaky spot.

As many floor slabs contain 2x3 or 2x4 sleepers, A,Illus.138, and wire reinforcing B, you will need an electric hammer, a hand saw and wire cutters. Rent the electric hammer and wire cutters.

Use extreme care when using an electric hammer. Wear safety goggles and only use the hammer after the rental store has shown you how. An electric hammer is an easy tool for an intelligent, alert person to use, but a dangerous tool for anyone who doesn't recognize its power.

Using a 1", or wider wood chisel, chisel a hole in one end of a piece of finished flooring. Make it large enough so you can drive a wrecking bar under the end, or side of an adjacent strip of flooring. You may have to chisel several holes in the same strip of flooring to get the first strip removed. Once you get started, you can save and reuse much of the flooring. When you have removed the finished flooring, and/or sub-flooring, start knocking a hole in the concrete some distance away from where you think the pipe is located. Study the feeder connections to radiators to make certain a feeder line isn't located in area selected. A feeder line is one that supplies each radiator. If possible, try to locate the heating or plumbing contractor who made the installation. He can save you considerable time, labor, mental anguish and money. By first tracing the line where it goes underground after leaving the boiler, you can frequently follow its exact direction by feeling the warmth or hearing the sound of water.

Proceed very cautiously until you break open a hole in concrete. Use care to break this hole clear of pipe. Use a hand hammer and chisel to chip hole larger until you can get your hand under slab. Dig earth out and keep feeling around until you find the pipe. Always use a hand hammer and chisel when you work near the pipe. Don't gamble using an electric hammer. One touch with an electric hammer and you need to replace more tubing than any leak normally requires.

When you find the leak, open up concrete so you or a plumber can cut out the leaking length of pipe and sweat two nipples and a length of tubing.

If you have to saw through a length of sleeper to make more work space, it's OK, and it doesn't need to be replaced.

Once you have located the pipe, make a chart showing distance from wall, its direction and depth below floor. Also indicate position of feeder lines.

Leave repaired pipe exposed until you run a two or three day test. When replacing fill, use only clean sand or dirt fill, and be sure it doesn't contain any nails, cinders or other foreign matters. Repatch concrete. Replace underlayment, etc., and say a prayer of thanks you could find and follow directions.

PLUMBING CHECK LIST FOR PROSPECTIVE BUYERS

The plumbing represents a big factor in evaluating the cost of a new home. The existing plumbing represents 10%, or more, of the price you pay for a house. It could represent as much as 20% if you have to replace it. For this reason, don't hesitate to spend time studying the condition of pipes and fixtures.

Note all exposed pipe joints. See if there's any discolorization caused by water leakage on walls or floor. Note whether there's any rust, moisture or leakage.

Turn each faucet on in kitchen and bathroom to see if you have as much pressure upstairs as down. See if there's enough pressure to have a downstairs faucet open while you draw water in an upstairs bathroom.

Taste the water. Is it to your liking? Note whether there is any odor, rust, scale, corrosion, or too much chlorine.

Flush each toilet. See how quickly water goes down drain. Flush it a second and third time to make certain there's no stoppage in a long waste line. Does the toilet work quietly, shut off as it should?

Check each sink, lavatory and tub. Place stopper in drain. Fill fixture and note how long it takes water to drain off.

Try the hot water line. Allow it to run a while to see how much hot water the system provides; how long it takes to "recover." Hot water heaters and tanks in many older homes were installed prior to dishwashers and washing machines. Many don't have the capacity to handle this equipment. In many cases the addition of only one piece of equipment requires a larger hot water tank.

If the house is served by a well, find out who has been servicing it. Get a written statement from owner concerning past water supply, and what he thinks its capacity is. Many wells fluctuate in capacity due to seasonal rains. The best time to judge a well's capacity is during, or after, a long drought, or at the end of a hot summer.

Are waste line clean out plugs easy to service? These should be placed in position shown, Illus. 8. Cleanout plugs should be located at least every 40 ft., never more than 50 ft. apart. Does a hot water heating system have an easy to reach drainage plug. Are waste lines to sink and lavatory undersize or do they meet local codes? A sink waste line should be 1-1/2" minimum. If line is longer than 40', it should be 2" minimum.

Does the water line chatter when you close a faucet quickly? Does a hot water heating line set up noises when supplying certain radiators?

If the house is heated with hot water lines embedded in a concrete slab, do rugs, or newly laid linoleum, or tile, cover a bad crack in concrete, or a ruptured pipe? It's always well to insist on turning the heat on, even during mid-summer, to test the system for noise, leaks, etc. Some boiler circulators, and/or pipes leak only occur when the system is operative.

Go up into attic and see whether there is, or was any leak around vent pipes going through roof. Water stains on framing, roof rafters, attic floors, etc., provide signs of previous leaks.

PLUMBING WALL

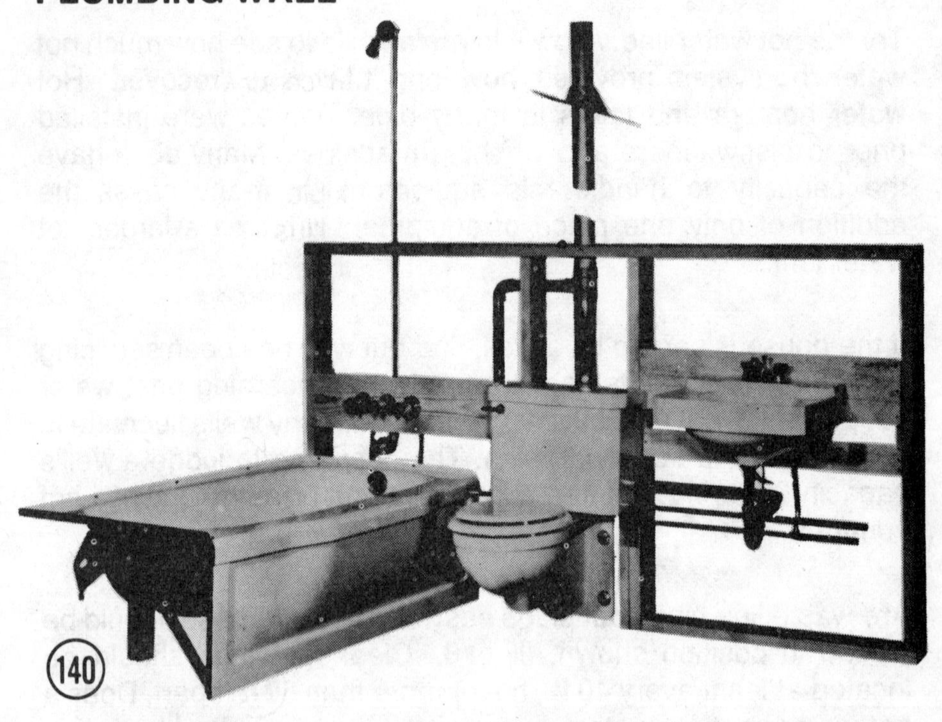

For those who want to install a bathroom quickly, and with as little labor as possible, the EB Plumbing Wall, Illus.140, provides an excellent solution.

No floor flange or closet bend is required when installing the EB Plumbing Wall, Illus. 140. Toilet waste empties into wall inlet, as does drainage from lavatory and bathtub. Since the plumbing wall is easy to install, select any location you desire. It is especially suited to one story houses built on a slab, and in two story installations when you don't want to disturb a ceiling below.

TELEPHONE SHOWER

One of the latest devices designed for better living is the "telephone" shower, Illus. 141. This easy to install, hand held shower permits washing any part without wetting others. It is particularly popular with women who want to shower without wetting their hair.

Those who don't have a bidet find this device almost as efficient in cleansing vital areas.

The telephone shower consists of a flexible hose that can be fastened directly to existing shower arm, Illus. 142. Or you can keep your shower and use the diverter connection, Illus. 143.

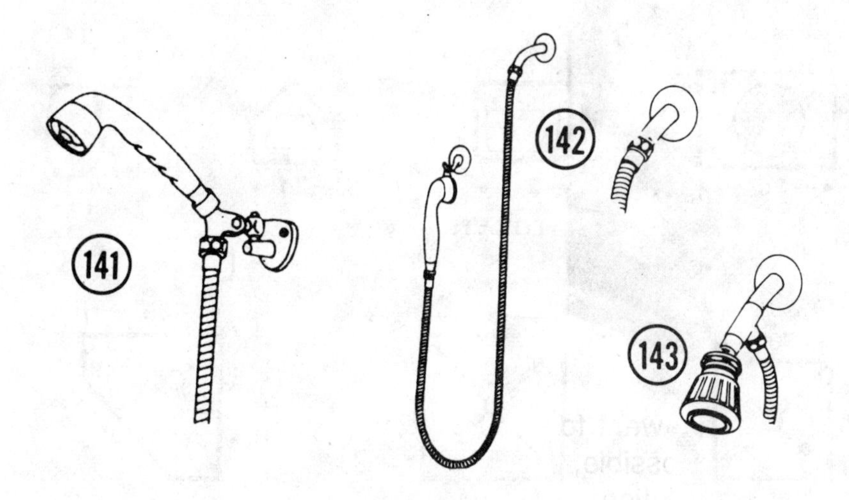

If you prefer to do away with showerhead and make a close fitting connection, a replacement elbow, Illus. 144, is also available. If your bath doesn't have a shower, remove discharge nozzle and replace with adapter spout, Illus. 145.

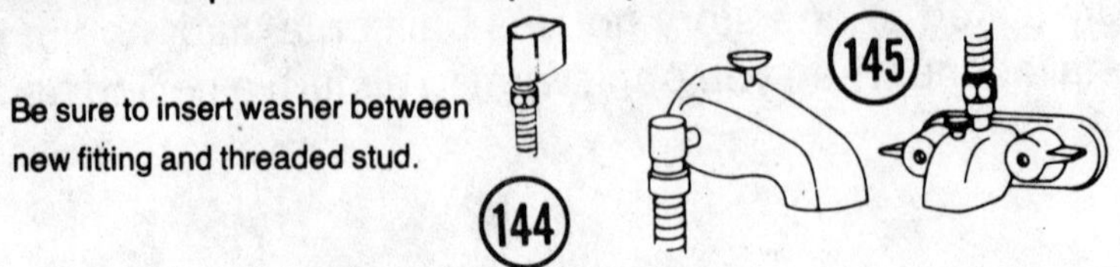

PLUMBING FIXTURE SIZES

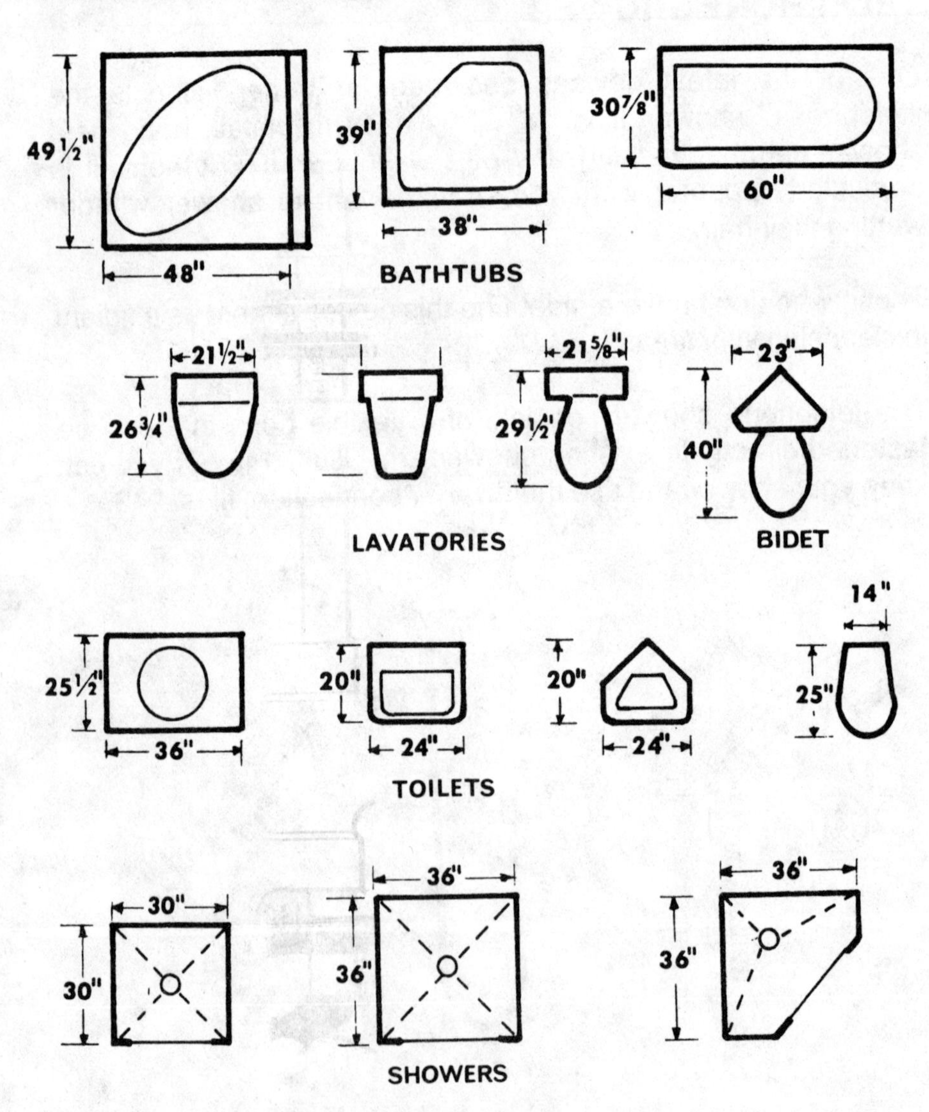

PLASTIC WATER, SEWER AND DRAIN PIPE

Plastic water, sewer and drain pipe is easy to install and requires no special tools or skill, Illus. 146. It's light in weight, easy to cut and a cinch to join. A 10 foot length of 4" pipe weighs less than 8 lbs. Hi-temperature plastic can be used for hot as well as cold water lines. It's perfect for septic systems, drainage lines to sewer or septic tanks, storm drains, septic system leaching fields, house and lawn drainage, etc. Since it's available in 10 and 20 ft. lengths, it requires fewer joints.

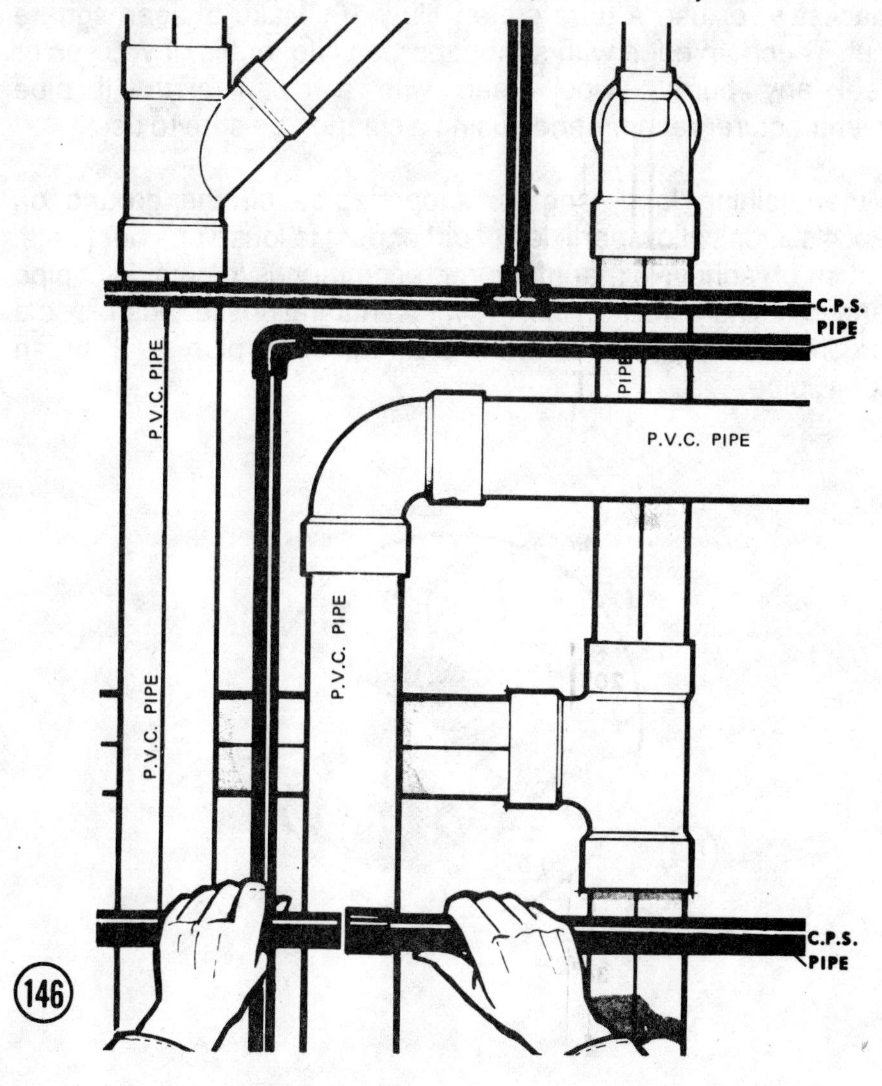

Always inspect trenches after digging to depth required to make certain there are no sharp rocks and no large stones above frost level that can heave a drainage line during a severe cold snap. Allow approximately 1/4" fall per foot for drainage except where codes specify other minimums.

Bury plastic pipe at least 24" below grade, and 36" under driveways.

To saw pipe to length required, use a mitre box and an ordinary hacksaw, or use a tube cutter, Illus. 10. Make a clean square cut. Touch up edge with sandpaper, a knife, or file, if you see or feel any burrs. Wipe clean with a clean cloth. If pipe manufacturer recommends using a cleaner, be sure to use it.

When joining long lengths, support pipe off the ground on 2 x 4's. Don't allow particles of dirt or dust to louse up your joints. Brush on adhesive manufacturer recommends to outside of pipe and to inside of fitting, Illus. 147 with a natural bristle brush. Use a brush that's half the size of the diameter of the pipe—a 2" brush for 4" pipe.

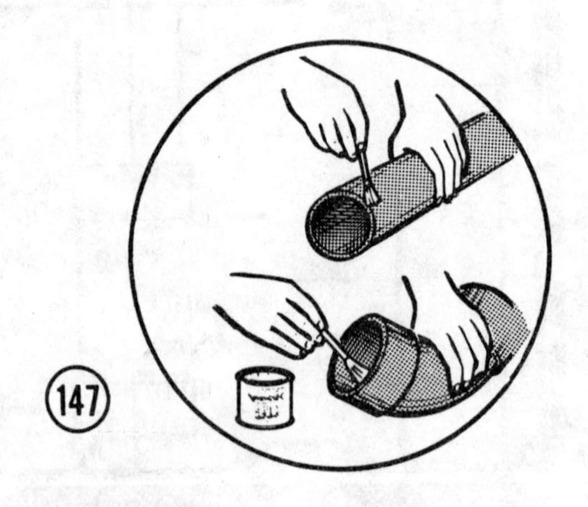

147

Insert pipe to full depth of fitting. Rotate pipe 1/4 turn, hold steady for 30 seconds (or length of time solvent manufacturer specifies) and allow it to set undisturbed for as long as directions on can specify. This can range from 3 minutes to half an hour, Illus. 148.

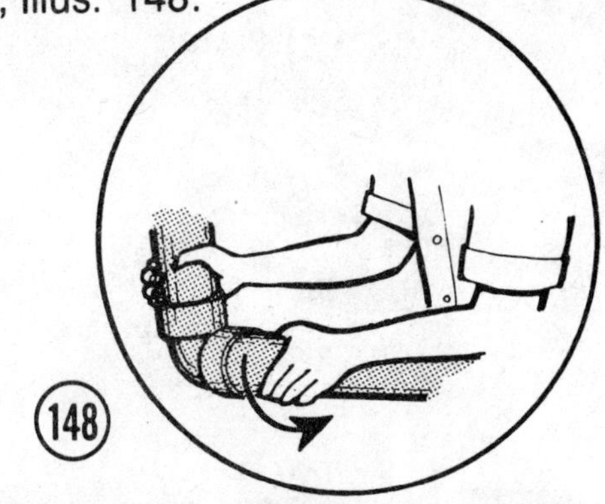

Some adhesive directions suggest joining pipe above ground, then immediately lowering into trench, and allow to set up time prescribed.

Illus. 149 shows the many different plastic fittings that are available.

There's nothing difficult about working with plastic pipe except to cut smoothly, eliminate burrs, clean up dust, apply solvent, and watch setting time. The solvent sets up fast. For this reason always make a dry run without solvent. When making up a joint where any change in direction is required, make a jig by placing 2 x 4 blocks in position needed to hold joint and straight lengths in exact position required.

Always use fittings manufactured by pipe manufacturer. This insures getting fittings that fit. Don't be snowed by price. A fitting that fits should be placed on a pipe dry. It should cling to pipe when held vertically. If it doesn't hold there is too much play in the fitting.

When making up an assembly containing pipe fittings, always bring the newest member to be added to the assembly, rather than move the assembly.

PLASTIC FITTINGS AVAILABLE

PIPE DIAMETER 2 – 3 – 4 – 6 in.

COUPLING

STRAIGHT

⅛ BEND

45 Deg. ANGLE

¼ BEND

90 Deg. ANGLE

WYE

45 Deg. ANGLE

SANITARY TEE

90 Deg. ANGLE

CAP

90 Deg. SWEEP

45 Deg. SWEEP

MALE ADAPTER
to N.P.T.

HUB ADAPTER TO
CLAY OR SOIL PIPE

SNAP COUPLING FOR
PERFORATED PIPE

CLAY SPIGOT ADAPTER

SPIGOT END OF CLAY PIPE

PLASTIC PIPE

PLASTIC FITTINGS CONTINUED

DOWNSPOUT ADAPTER

For 2" x 3" leader; also available for 3" x 4" leader.

REDUCING COUPLING

(each coupling reduces one size)

PIPE DIAMETER 3"-4"-6"

ADAPTER FOR FIBER PIPE

FIBER PIPE

PLASTIC PIPE

CLEAN-OUT BUSHING

PIPE DIAMETER 3"-4"

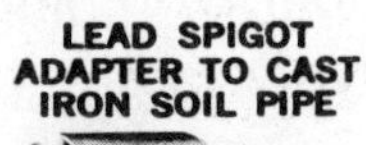

LEAD SPIGOT ADAPTER TO CAST IRON SOIL PIPE

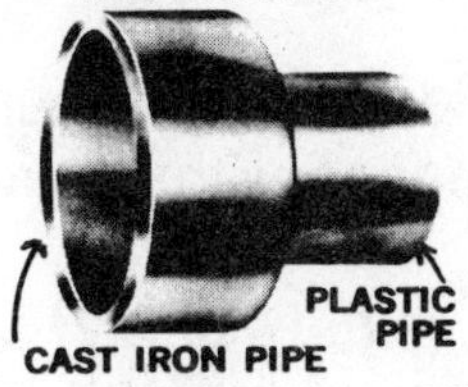

CROSS

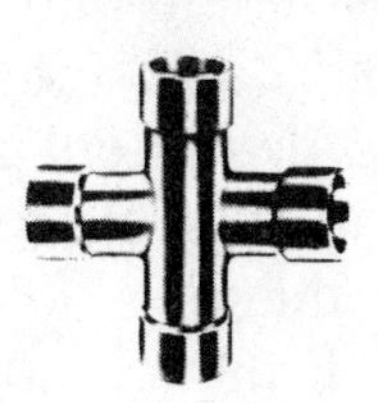

PIPE DIAMETER 4"

THREADED PLUG

150

When assembling threaded plastic pipe always use a strap wrench, Illus.151, rather than a pipe wrench. Never use pliers. Don't overtighten a fitting. Threaded plastic pipe is always joined hand tight plus one to one and a half turns. To make certain you know how many turns are required, dry fit pipe and fitting. Count the number of turns needed to hand tighten. Remove fitting from pipe. Apply tape or dope to male thread only.

151

Strap Wrench

Screw fitting on to pipe using the same number of turns as previously required. Now turn the fitting, using a strap wrench, one to one and a half turns beyond the hand tightened point.

CAUTION. Keep solvent cans covered when not in use. Don't use welding solvent compounds after they begin to gel or thicken. Do not add thinners. Throw it away and get a new can of the welding solvent pipe manufacturer recommends. Never attempt to assembly pipe in freezing or below freezing temperature or when pipe or fitting is wet.

PLASTIC HIGH TEMPERATURE PIPE

You can make professional plumbing repairs to hot and cold water lines with high temperature plastic pipe if you follow certain basic rules.

1. Always cut pipe clean, remove burrs, clean pipe and fitting with a cloth dampened with cleaner manufacturer recommends.
2. Brush on solvent, join, twist and allow to set as described previously.
3. Support pipe every 32" with clamps recommended by plastic pipe manufacturer. These permit pipe to expand with heat, contract with cold. Never clamp plastic, copper or any material that expands and contracts too securely. You must allow it to move freely.

Illus. 152 shows how high temperature plastic can be used on cold and hot water lines.

TRANSITION ADAPTER

(Note: Do not enclose transition fittings in walls or other inaccessible places as a matter of good plumbing practice.)

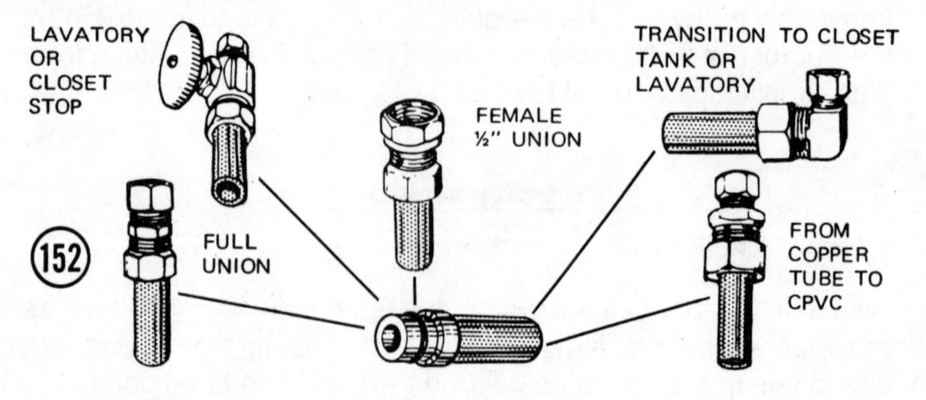

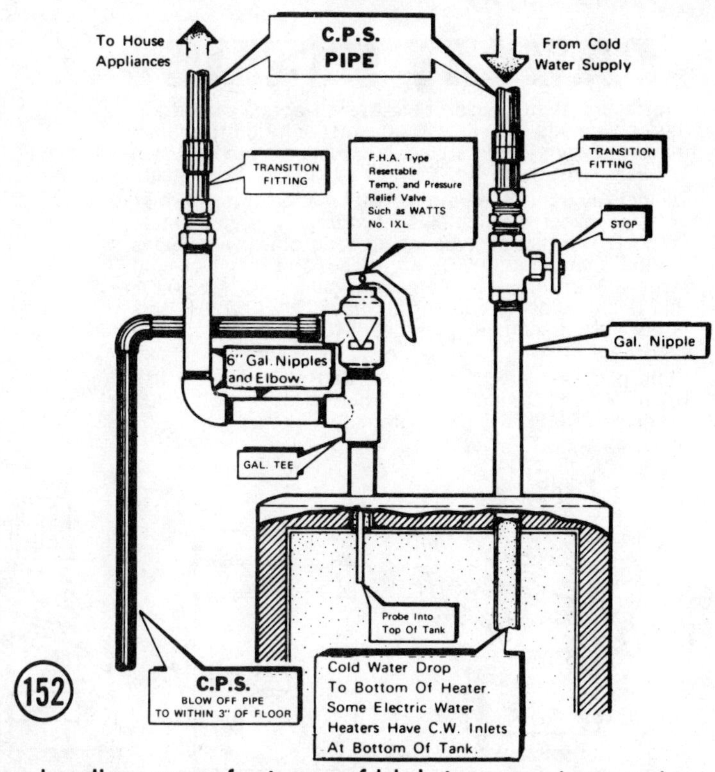

One leading manufacturer of high temperature water pipe offers color coded pipe. The hot water side can be orange, the cold water green. Illus. 152 shows how galvanized nipples, elbows and tee are installed in a hot water heater then connected to high temperature plastic.

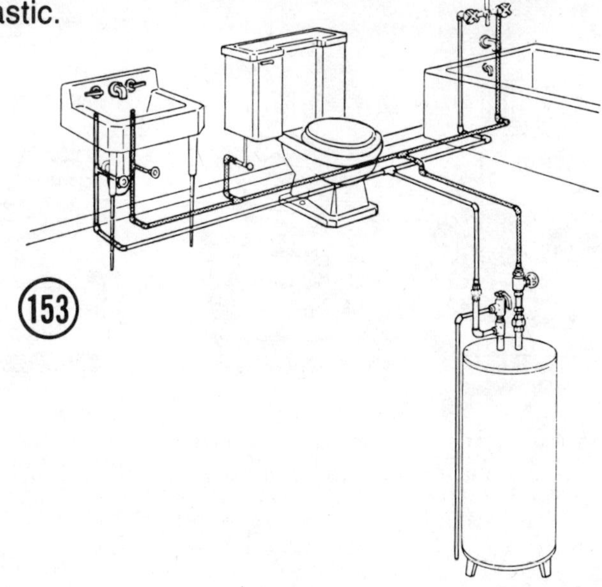

FAUCET STEMS

The faucet stems shown here are all actual size drawings, giving you the easiest, quickest solution to replacing old or worn faucet stems. All stems are grouped into 12 basic sizes, from No. 1 (smallest) to No. 12 (largest).

To identify your stem, simply place it on the drawings until an exact duplicate is found. This is your stem number. The text accompanying each drawing includes original manufacturer, OEM part no. and if hot or cold sides are different. If number includes "C" (FS1-9C), this is cold side. If number includes "H" (FS1-9H), this is hot side. If number includes "HC" (FS1-13HC), stem fits either side.

Stem packages are color-coded: Red for HOT side (H); Green for COLD side (C); Brown for Hot or Cold (HC). Order by COMPLETE part number.

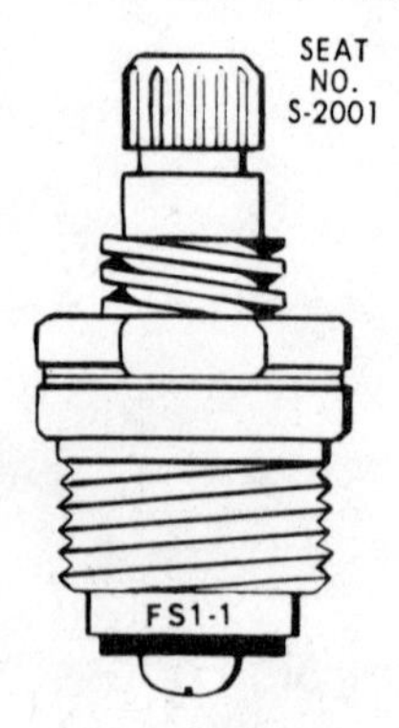

FS1-1C **FS1-1H**
Fits AMER. BRASS, EMPIRE BRASS, STREAMWAY
118 Concealed Sink Faucet

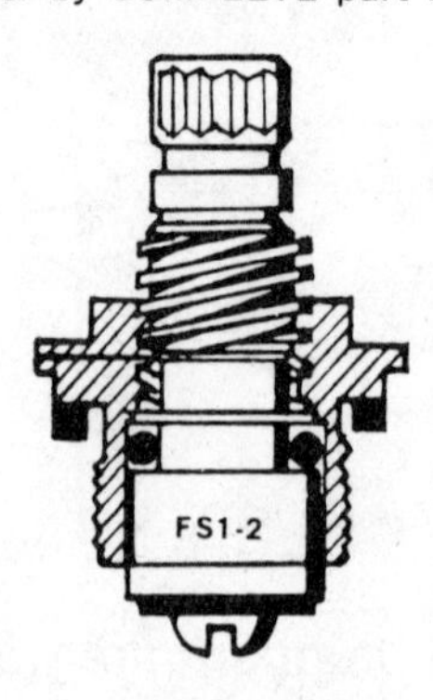

FS1-2C **FS1-2H**
Fits AMER. BRASS, EMPIRE BRASS, STREAMWAY: 104, 1104, 105, 1105 Sink

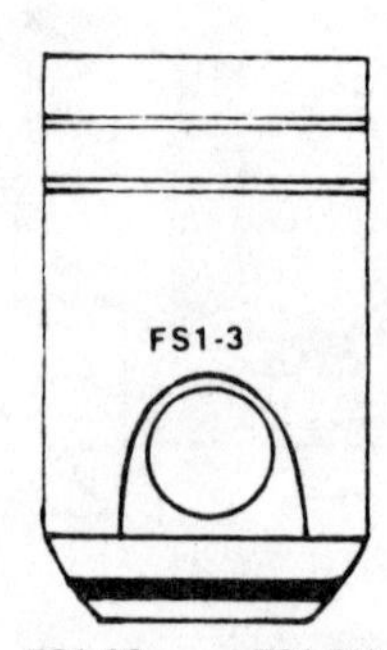

FS1-3C **FS1-3H**
Fits AMER. STANDARD (Re-Nu Barrel)
OEM 20336-08 RH, 20563-08 LH

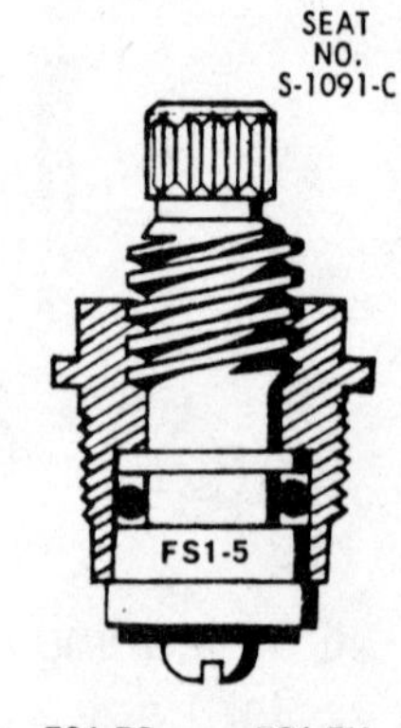

FS1-5C **FS1-5H**
Fits CENTRAL SU-357-K
Model Lav., Bath, Laundry and Sink

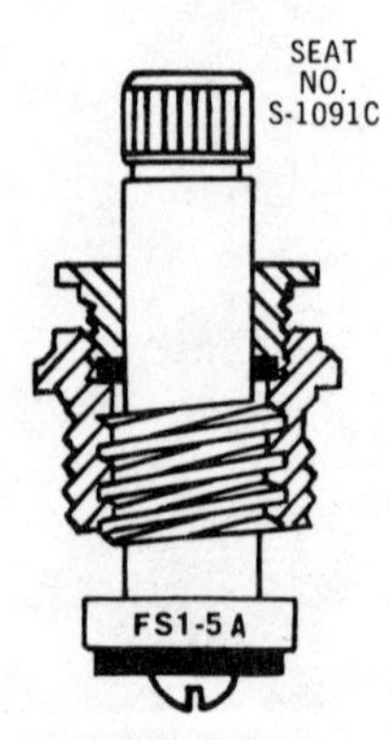

FS1-5AH **FS1-5AC**
For all bath, lavatory, laundry and sink faucets.
Interchanges with FS1-5.

FS1-6C **FS1-6H**
Fits GERBER
OEM RPA-28-1, 29-1

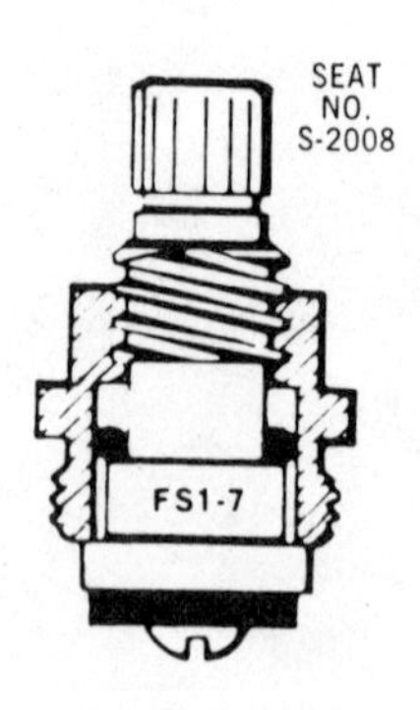

FS1-7C **FS1-7H**
Fits INDIANA BRASS
OEM 631-C & D

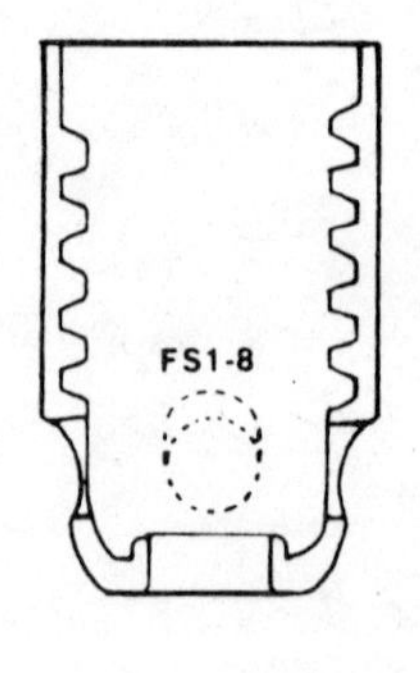

FS1-8C **FS1-8H**
Fits KOHLER
OEM 32462 RH, OEM 39717 LH

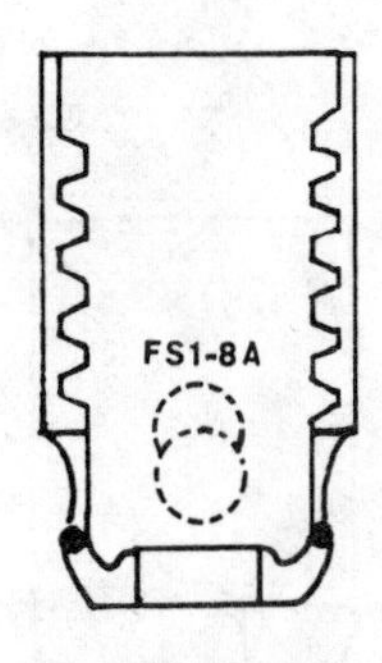

FS1-8AC FS1-8AH
Same as OEM 32462
RH (FS1-8C)
Same as OEM 39717
LH (FS1-8H)
but with "O" Ring

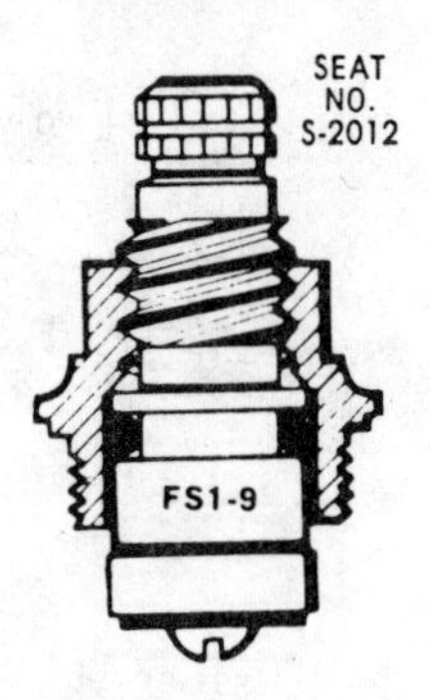

FS1-9C FS1-9H
Fits PRICE-PFISTER
635-645

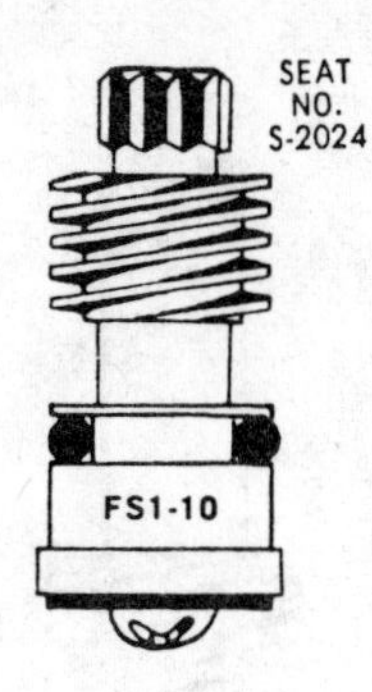

FS1-10C FS1-10H
Fits REPCAL
OEM 14147 RH,
OEM 14148 LH
Old No. 16-5

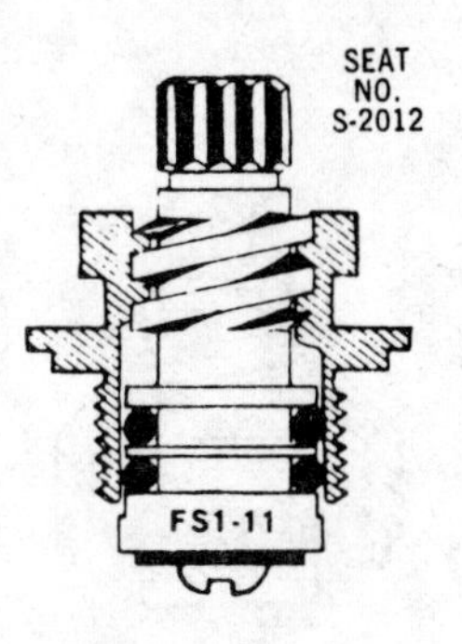

FS1-11C FS1-11H
Fits SAYCO-NOLAND
OEM 45-D-12 Stem,
57-D-13 Bonnet for
280 Fitting

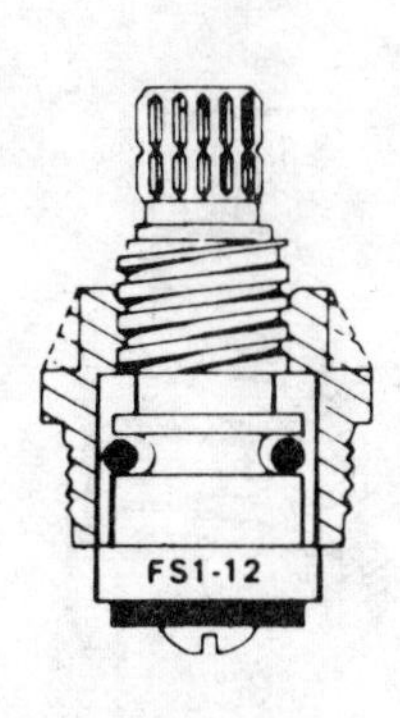

FS1-12C FS1-12H
Fits UNION BRASS
OEM 1837-A

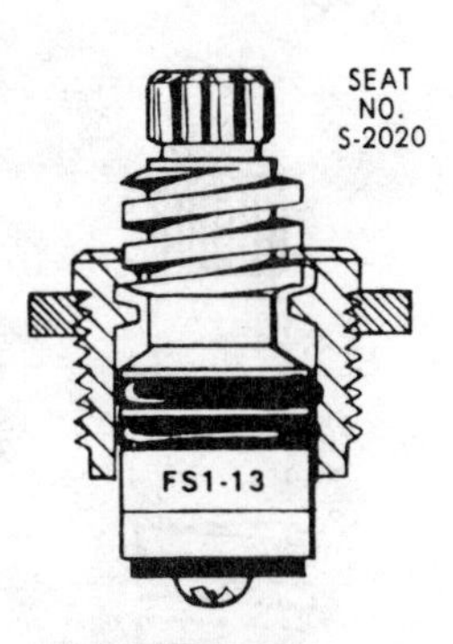

FS1-13HC
Fits HARCRAFT
OEM 3247:
10-100 through 10-171
and 12-300 through
12-302

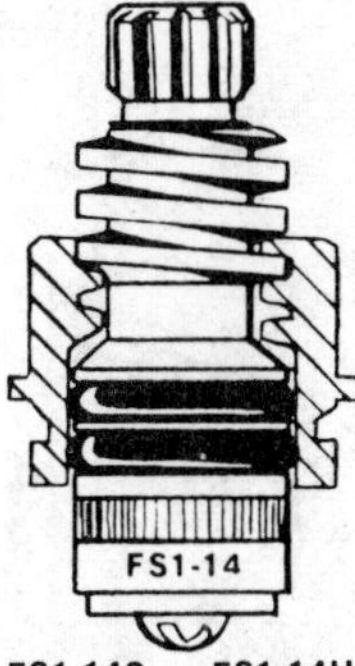

FS1-14C FS1-14H
Fits HARCRAFT OEM
99-3128-H, 99-3129-C:
A-50, A-51A, A-52A,
A-160, A-180, A-190,
A-540, A-541 Swing
Spouts, Center Sets,
Exp. Ledge Types

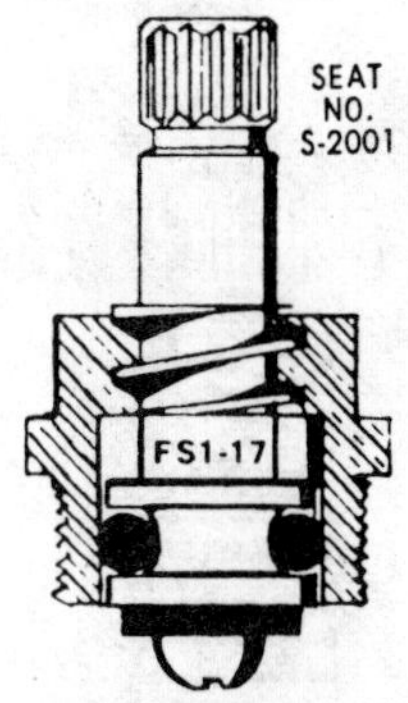

FS1-17C FS1-17H
Fits STERLING
OEM 99S-8079-H,
99S-8080-C:
20-310 Series
Lav. Fittings

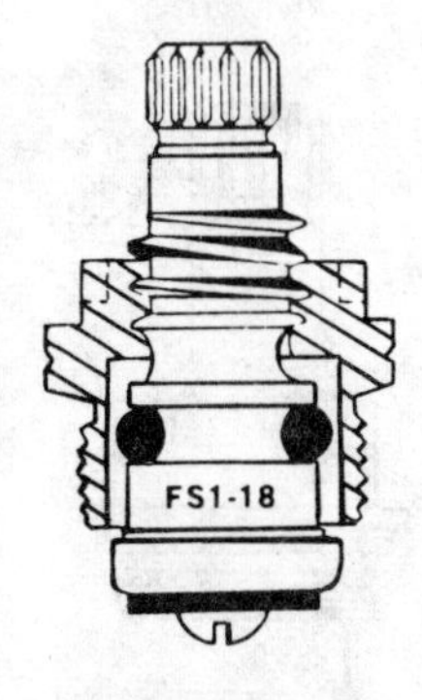

FS1-18C FS1-18H
Fits STERLING
OEM 99S-8109-H,
99S-8110-C

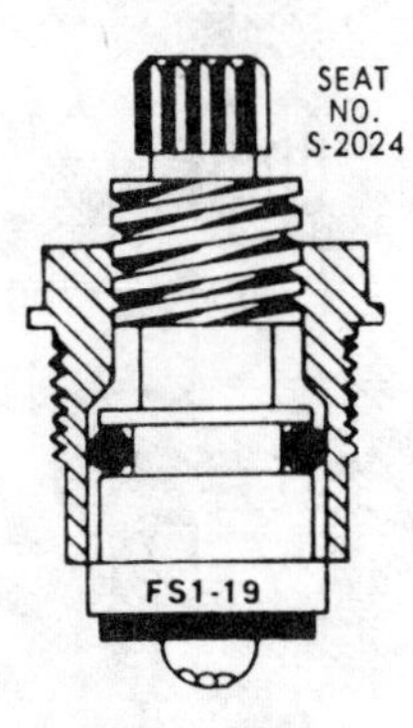

FS1-19C FS1-19H
Fits REPCAL
OEM FB 9173 RH,
OEM FB 9174 LH
Also 16-3A

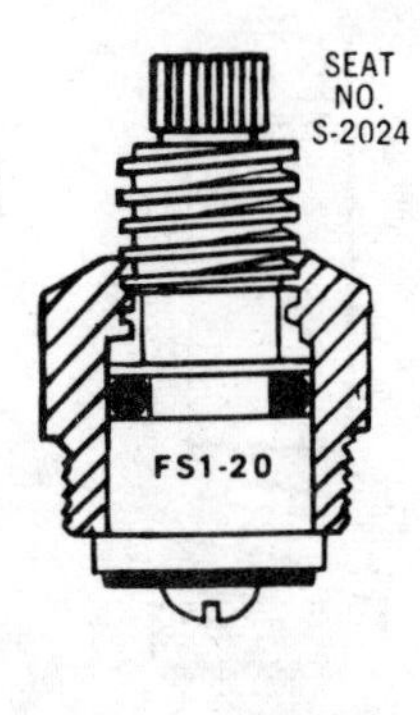

FS1-20C FS1-20H
Fits CRANE Crestmont
REPCAL, Riviera
OEM FB9137

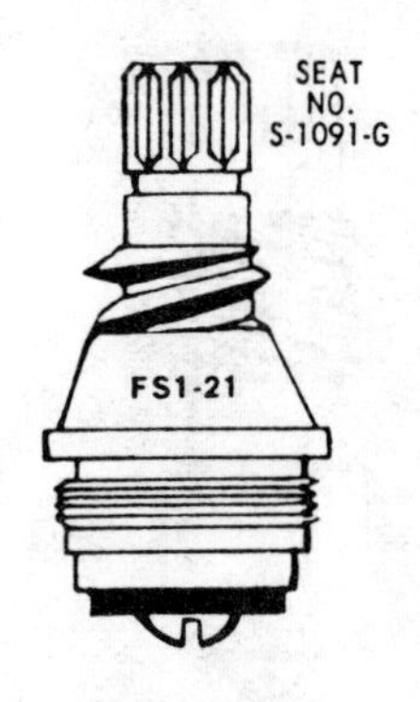

FS1-21HC
Fits BRIGGS
OEM 22268 Stem,
22269 Bonnet,
T-8832-33
Trim Line Lav.

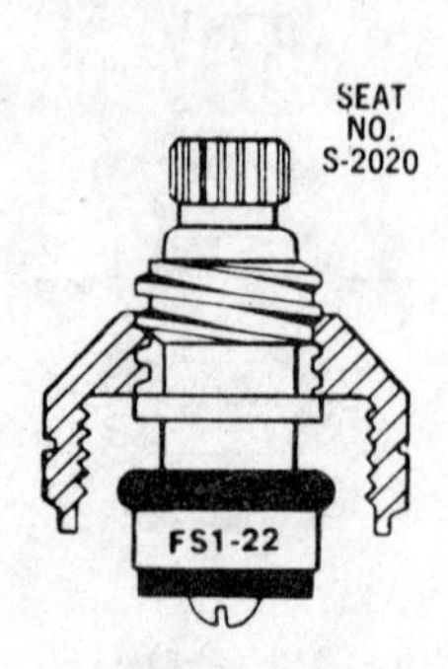

FS1-22C FS1-22H
Fits ELJER
OEM 5288-1-RH,
5288-2-LH:
No. 4 Unit

FS1-23C FS1-23H
Fits AMER. STANDARD
OEM 64804-07-RH,
50637-07-LH,
Fits C530-560 Colony
Lav. Fitting

FS1-61C FS1-61H
Fits GERBER
OEM 29-1: 250

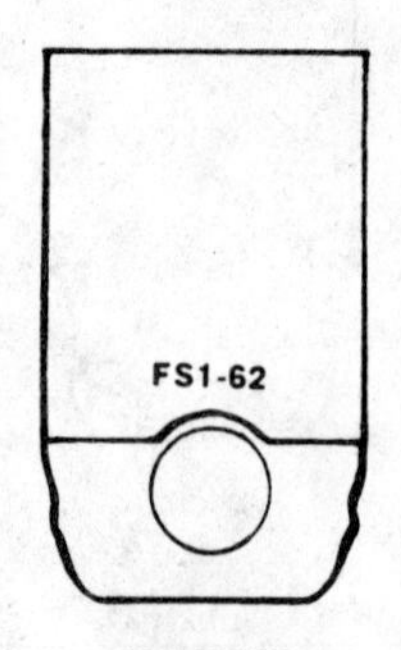

FS1-62HC
Fits KOHLER
OEM 32473
(Renew Barrel)
No Thread

FS1-63C FS1-63H
Fits SAYCO-NOLAND
OEM 45-D-12
for 280 Fitting

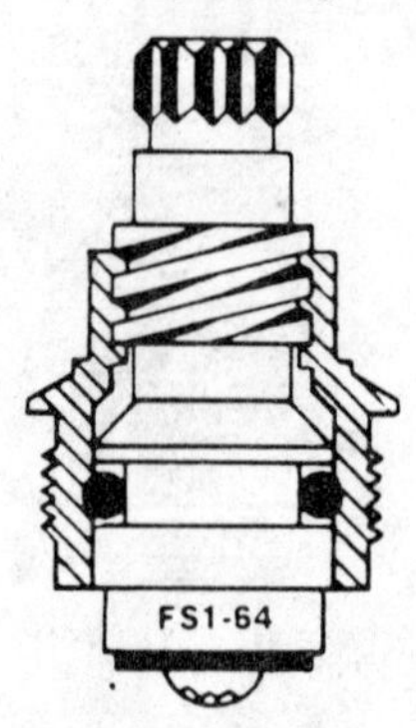

FS1-64C FS1-64H
Fits SEARS
Universal-Rundle Lav.

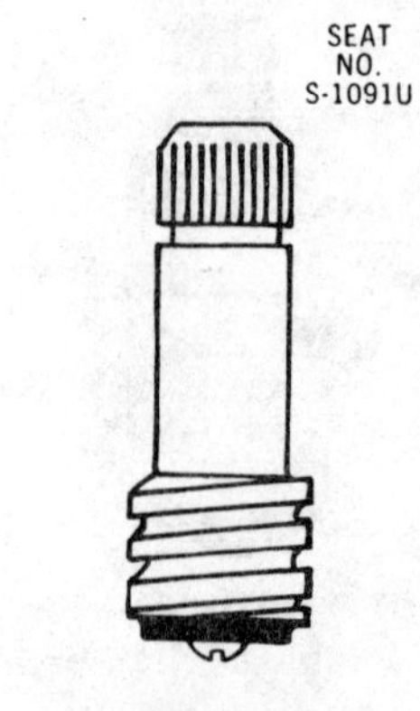

FS1-65C FS1-65H
Fits SPEAKMAN KENT
Lavatory and Sink
Fittings Models
SK-540-41, SK-560-61,
SK-570-71

FS2-1C FS2-1H
Fits AMER. STANDARD
OEM 6079-04-H,
6080-04-C:
R-4100-1-2-3

FS2-2C FS2-2H
Fits AMER. STANDARD
OEM 853-14-R,
856-14-L:
R-4100-1-2-3

FS2-3C FS2-3H
Fits AMER. STANDARD
OEM 54156-04-H,
54157-04-C;
CROSLEY C-F-100-200
SCHAIBLE 922-26,
SCH-1622;
YOUNGSTOWN-MULLINS
5505, 60954-55

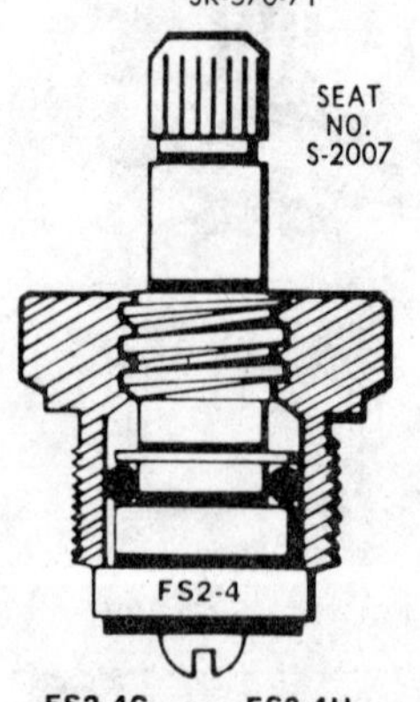

FS2-4C FS2-4H
Fits GERBER
260, 225-40, 53,
OEM RPA 28-2, 29-2

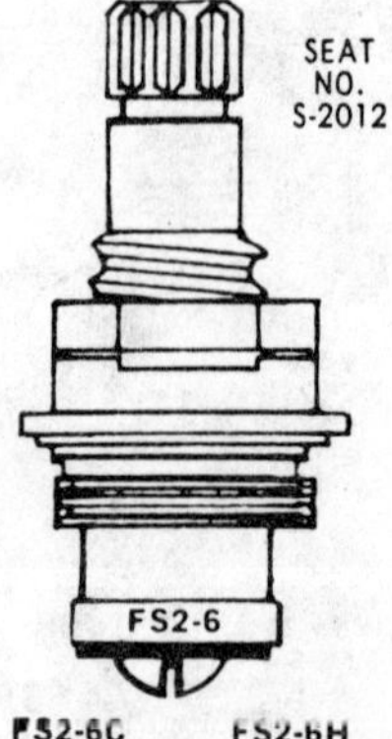

FS2-6C FS2-6H
Fits PRICE-PFISTER
OEM 03156-01-H,
03156-02-C

FS2-7HC
Fits PRICE-PFISTER
OEM 3147
460-61 Single
Lav. Faucet

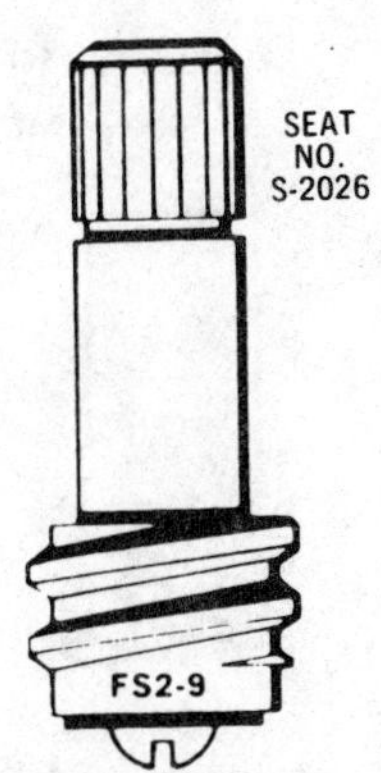

FS2-9C FS2-9H
Fits SPEAKMAN
OEM 3-336-C, 3-337-H:
3-292 (G 3-181) C,
3-291 (G 3-180) H,
SK-561

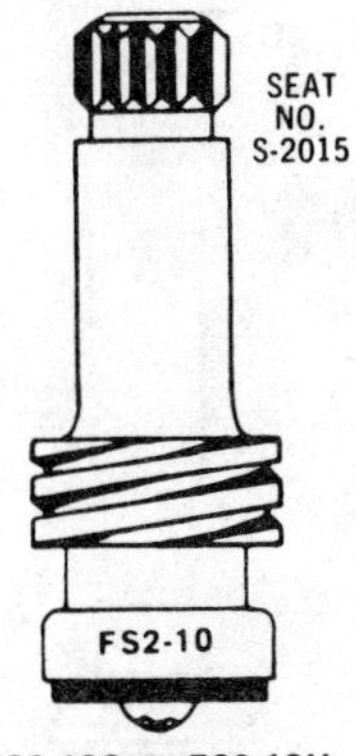

FS2-10C FS2-10H
Fits SEARS Homart

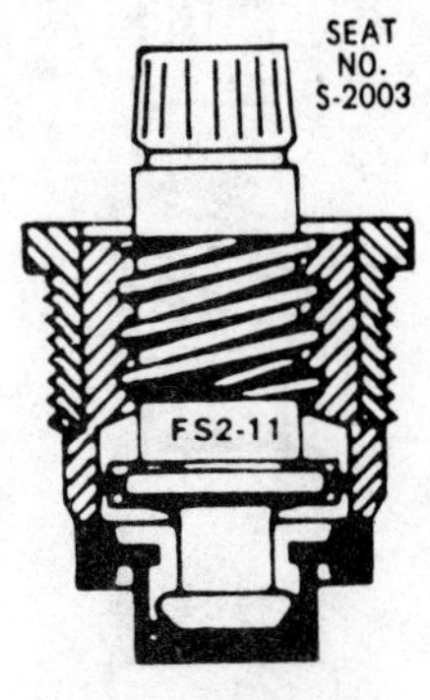

FS2-11C FS2-11H
Fits
AMERICAN STANDARD
OEM 72950-07-RH,
72951-07-LH

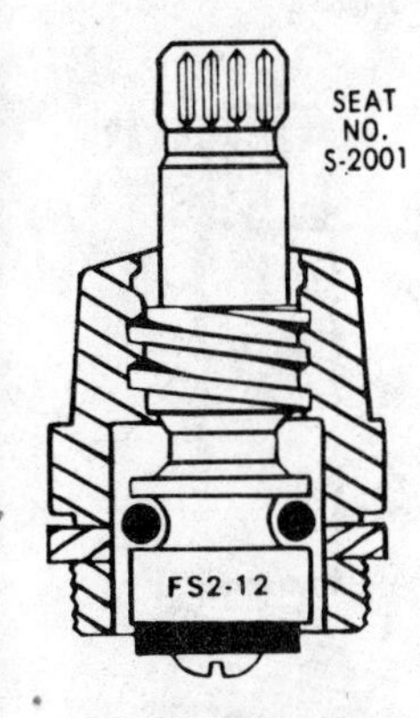

FS2-12C FS2-12H
Fits STERLING
OEM 99S-8153-H,
99S-8154-C

FS2-14HC
Fits PRICE-PFISTER
OEM 03175-01-00Z:
43-010 through
43-124 Fittings

FS2-61C FS2-61H
Fits GERBER
OEM 29-2:
260, 225-40, 53

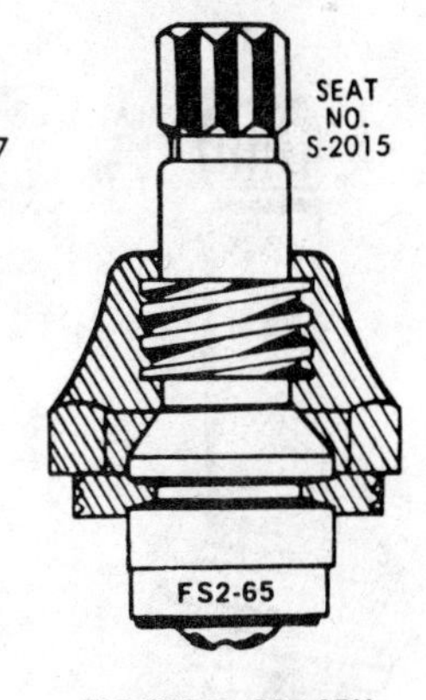

FS2-65C FS2-65H
Fits SEARS-MILWAUKEE
OEM 5077 RH, 5078 LH

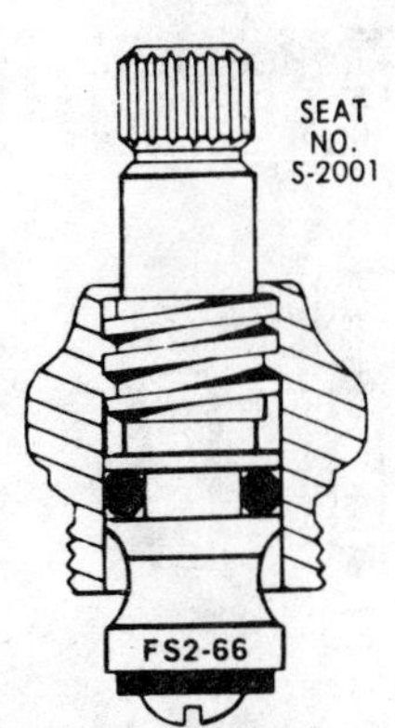

FS2-66HC
Fits STERLING
OEM 99S-8131-H,
99S-8132-C

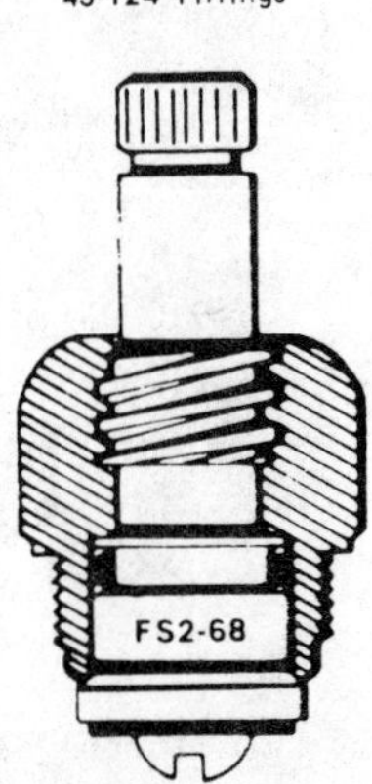

FS2-68C FS2-68H
Fits UNION BRASS-
GOPHER-ST. PAUL
OEM 3402,
1837-A; L-7370

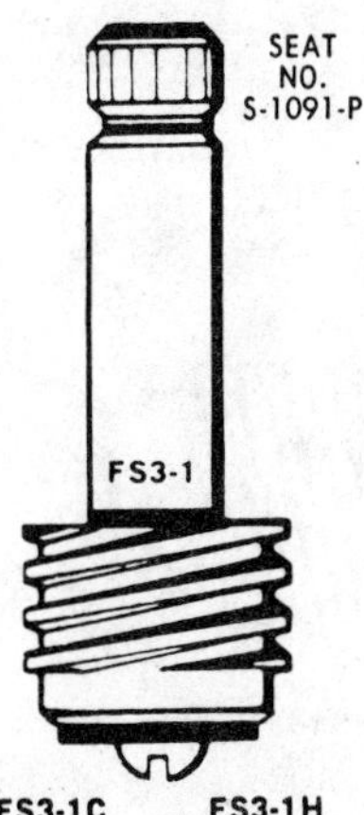

FS3-1C FS3-1H
Fits
AMERICAN KITCHENS
OEM 5771-C, 5772-H:
LF-110, 120, 210

FS3-2C FS3-2H
Fits AMER. STANDARD
OEM 7617-07-LH,
7616-07-RH:
B-876-8

FS3-3C FS3-3H
Fits AMER. STANDARD
OEM 711-17-LH,
712-27-RH:
B-876-8

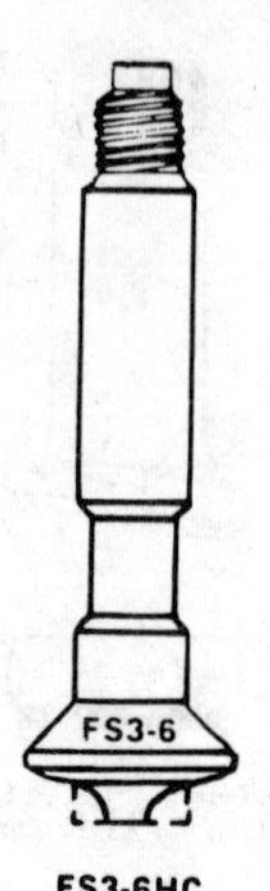

FS3-6HC
Fits CRANE
OEM F12537
for
Magic-Close Faucet

FS3-7C FS3-7H
Fits KOHLER
OEM 34070-C, 34071-H:
K-8610-A,
K-8005A-06-09

FS3-9C FS3-9H
Fits SPEAKMAN
OEM G3-158H, G3-159C:
S-4760-61, S-4770-71,
S-4700

FS3-11C FS3-11H
Fits UNION BRASS-
OEM P-107-H, P-107-C:
L-7335-41, L-7315-21

FS3-13C FS3-13H
Fits DICK BROS.
OEM 2011-H, 2011-C:
D-4008-10
Wall Faucet

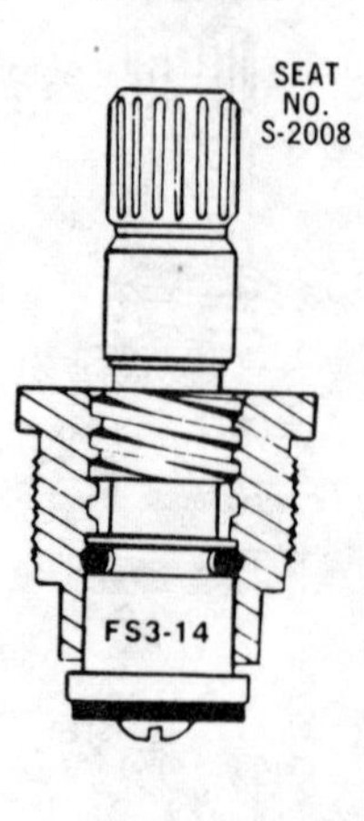

FS3-14C FS3-14H
Fits
MILWAUKEE FAUCET
OEM 5012-H, 5012-C:
Fits K-4009 Lav.
Center Set

FS3-15C FS3-15H
Fits CENTRAL BRASS
OEM SU-1855-47 R or L;
21, 23, 26, 28, 21-S,
26-S, 47½, 48½

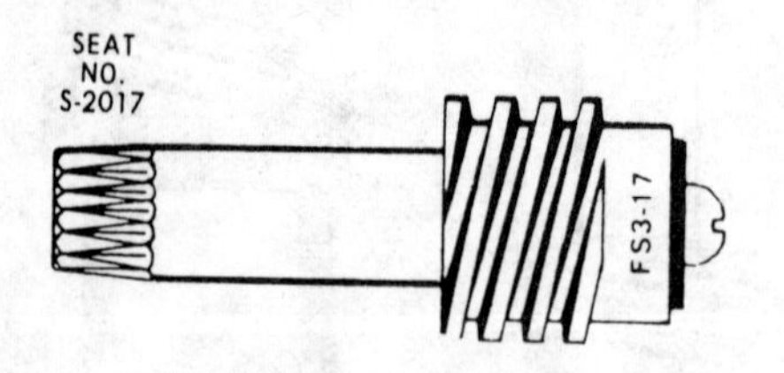

FS3-17C FS3-17H
Fits AMER. STANDARD
OEM 25535-02-RH,
25536-02-LH:
TL 30-37 Tract Line
Lav. Fitting

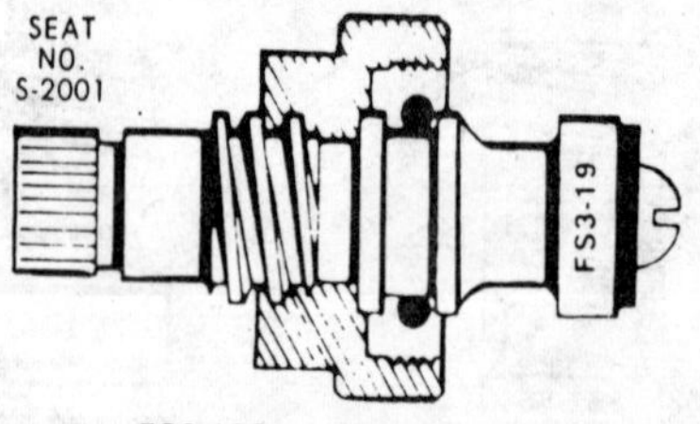

FS3-19C FS3-19H
Fits STERLING
OEM 99S-8126C,
99S-8125-H:
S-1100 (New)

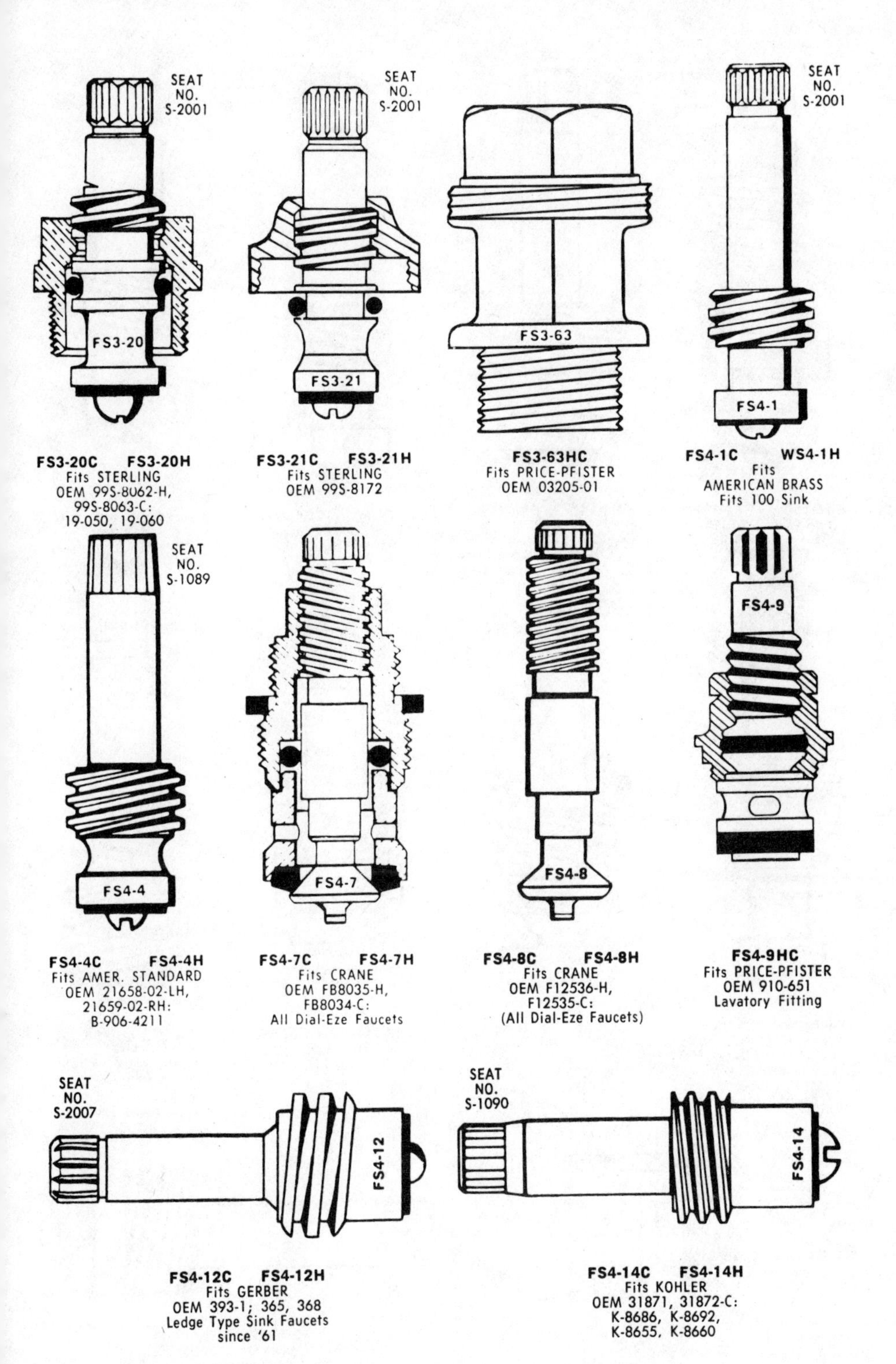

FS3-20C FS3-20H
Fits STERLING
OEM 99S-8062-H,
99S-8063-C:
19-050, 19-060

FS3-21C FS3-21H
Fits STERLING
OEM 99S-8172

FS3-63HC
Fits PRICE-PFISTER
OEM 03205-01

FS4-1C WS4-1H
Fits
AMERICAN BRASS
Fits 100 Sink

FS4-4C FS4-4H
Fits AMER. STANDARD
OEM 21658-02-LH,
21659-02-RH:
B-906-4211

FS4-7C FS4-7H
Fits CRANE
OEM FB8035-H,
FB8034-C:
All Dial-Eze Faucets

FS4-8C FS4-8H
Fits CRANE
OEM F12536-H,
F12535-C:
(All Dial-Eze Faucets)

FS4-9HC
Fits PRICE-PFISTER
OEM 910-651
Lavatory Fitting

FS4-12C FS4-12H
Fits GERBER
OEM 393-1; 365, 368
Ledge Type Sink Faucets
since '61

FS4-14C FS4-14H
Fits KOHLER
OEM 31871, 31872-C:
K-8686, K-8692,
K-8655, K-8660

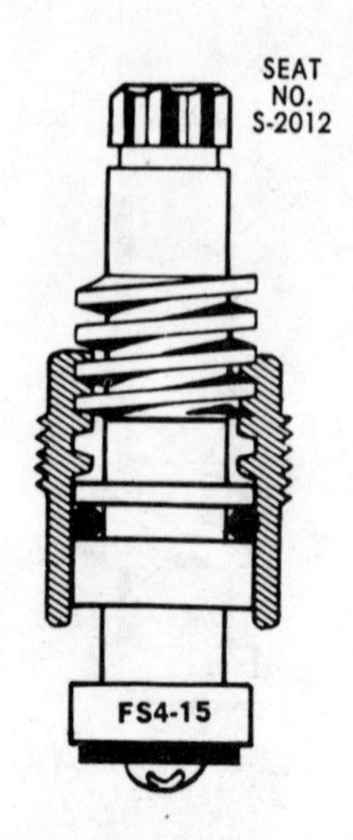

FS4-15C FS4-15H
Fits PRICE-PFISTER
OEM 03151-01-RH,
03151-02-LH

FS4-16C FS4-16H
Fits REPCAL
OEM F14143-H,
F14145-C; 13-511
205 Wall Sink
123 Laundry

FS4-21C FS4-21H
Fits STERLING
OEM 99S-0064-H,
99S-0294-C; S-100

FS4-22C FS4-22H
Fits STERLING
OEM 99S-0148-H,
99S-0648-C:
S-200, S-1120-22

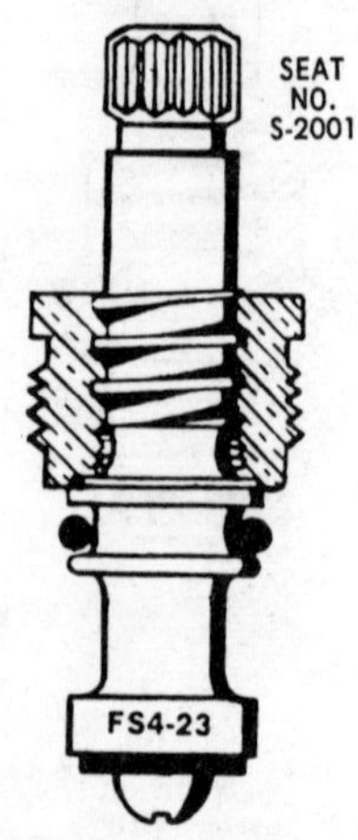

FS4-23C FS4-23H
Fits STERLING
OEM 99S-8021-H,
99S-8020-C:
20-10 (N300)
Lav. Fitting

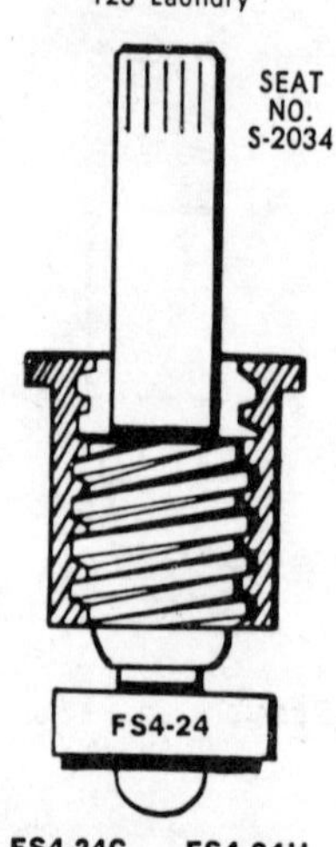

FS4-24C FS4-24H
Fits CRANE New Sleeve
Unit (25A-25.3/8):
C-32180-85-86,
C-32165-66-67 Lav.
C-32279-81-82 Lav.
C-32835 Sink,
C-32746-S Wall Sink

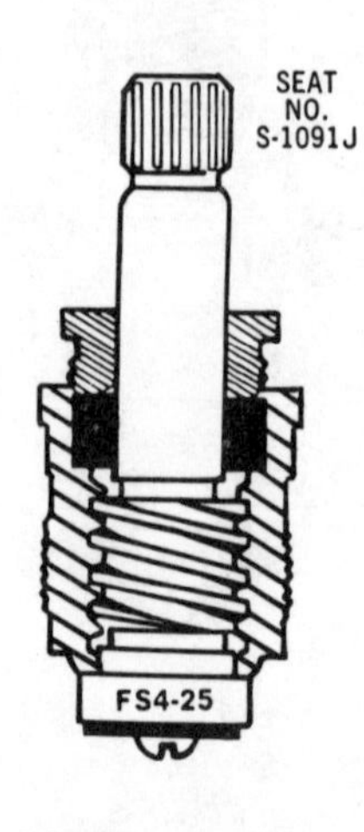

FS4-25C FS4-25H
Fits ELJER
OEM 4788 No. 3 Unit

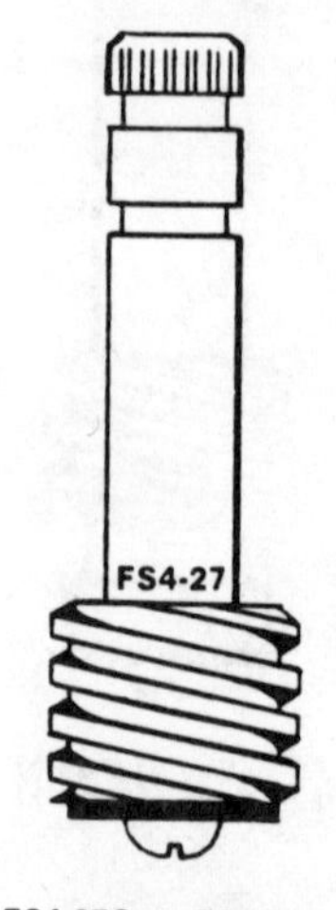

FS4-27C FS4-27H
Fits
AMERICAN STANDARD
OEM 64703-07 RH,
50668-07 LH
Colony Trim

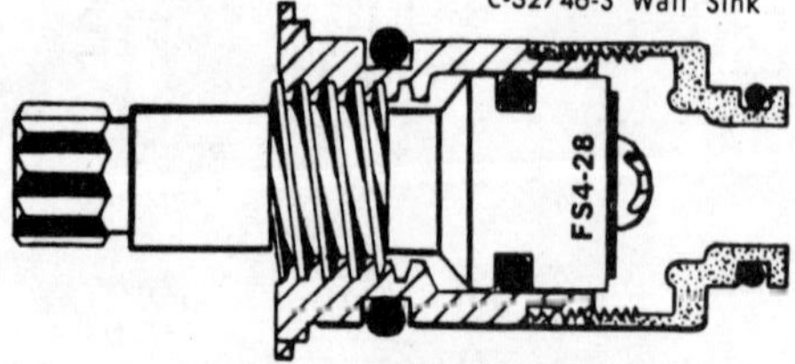

FS4-28C FS4-28H
Fits SEARS, ELKAY,
UNIVERSAL-RUNDLE
OEM P-1077, 32008-L

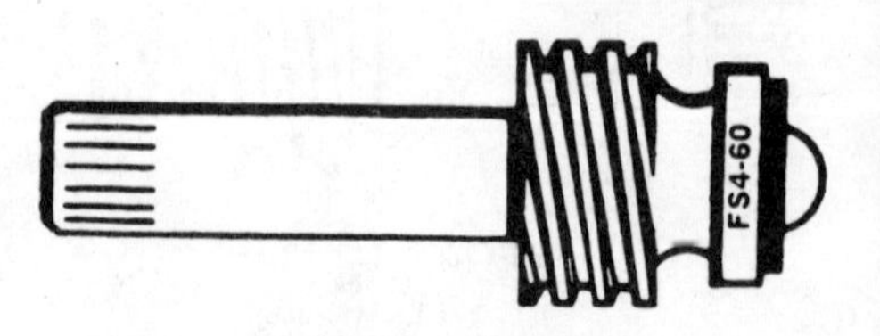

FS4-60C FS4-60H
Fits DICK BROS.
3057 Deck Faucet

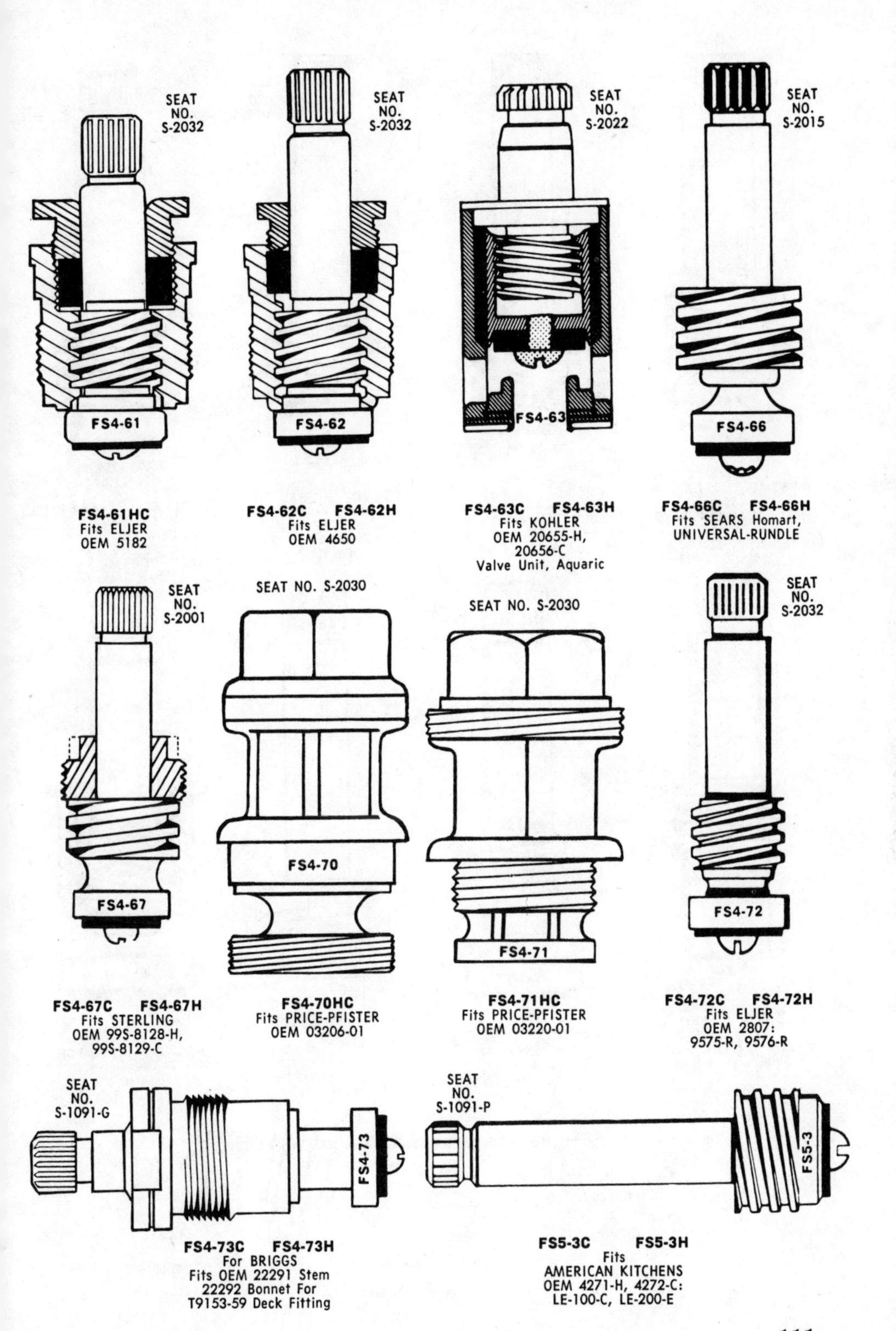
SEAT
NO.
S-2032
FS4-61
SEAT
NO.
S-2032
FS4-62
SEAT
NO.
S-2022
FS4-63
SEAT
NO.
S-2015
FS4-66
FS4-61HC
Fits ELJER
OEM 5182
FS4-62C FS4-62H
Fits ELJER
OEM 4650
FS4-63C FS4-63H
Fits KOHLER
OEM 20655-H,
20656-C
Valve Unit, Aquaric
FS4-66C FS4-66H
Fits SEARS Homart,
UNIVERSAL-RUNDLE
SEAT
NO.
S-2001
FS4-67
SEAT NO. S-2030
FS4-70
SEAT NO. S-2030
FS4-71
SEAT
NO.
S-2032
FS4-72
FS4-67C FS4-67H
Fits STERLING
OEM 99S-8128-H,
99S-8129-C
FS4-70HC
Fits PRICE-PFISTER
OEM 03206-01
FS4-71HC
Fits PRICE-PFISTER
OEM 03220-01
FS4-72C FS4-72H
Fits ELJER
OEM 2807:
9575-R, 9576-R
SEAT
NO.
S-1091-G
FS4-73
SEAT
NO.
S-1091-P
FS5-3
FS4-73C FS4-73H
For BRIGGS
Fits OEM 22291 Stem
22292 Bonnet For
T9153-59 Deck Fitting
FS5-3C FS5-3H
Fits
AMERICAN KITCHENS
OEM 4271-H, 4272-C:
LE-100-C, LE-200-E

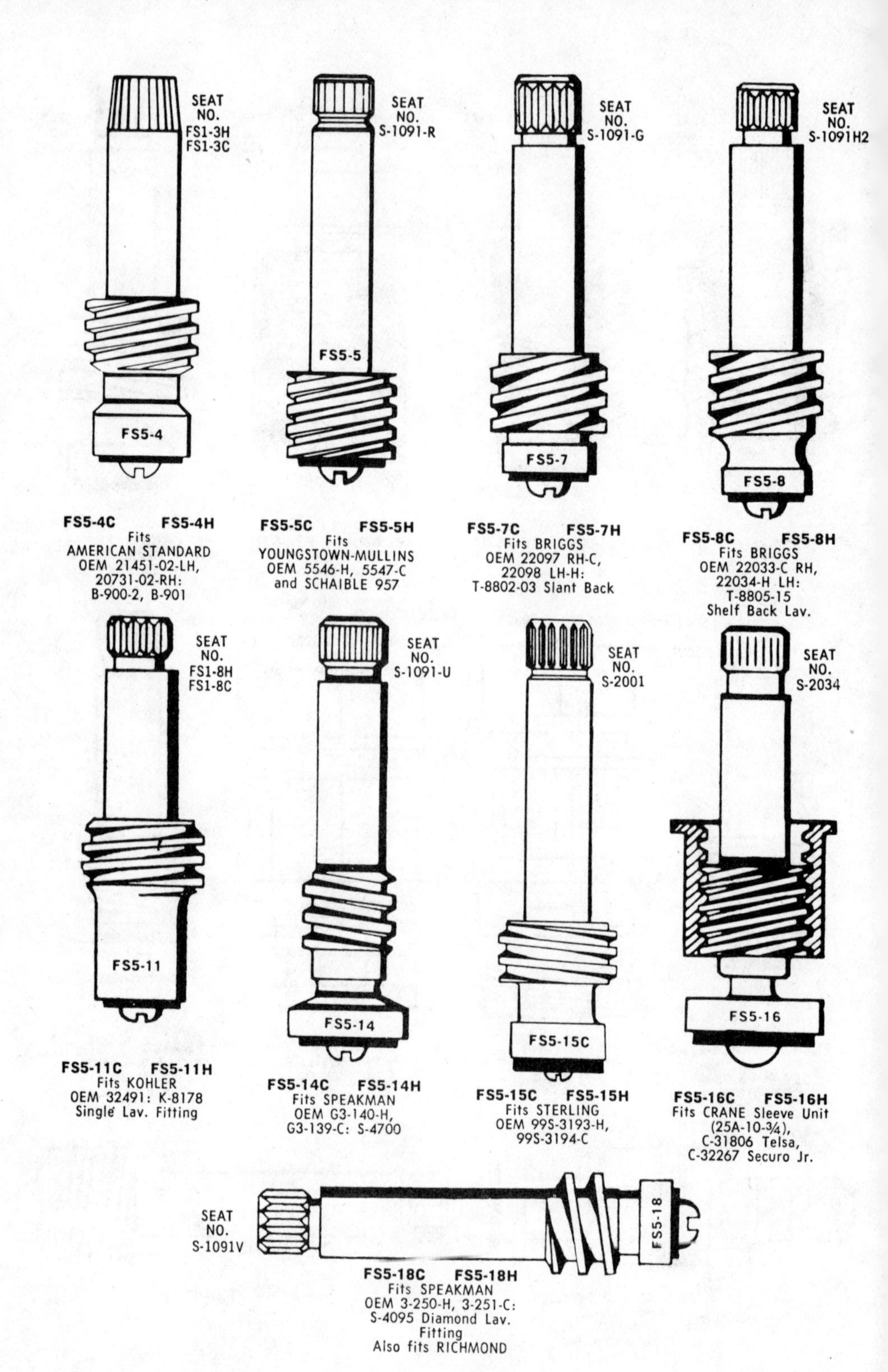

FS5-4C **FS5-4H**
Fits
AMERICAN STANDARD
OEM 21451-02-LH,
20731-02-RH:
B-900-2, B-901

FS5-5C **FS5-5H**
Fits
YOUNGSTOWN-MULLINS
OEM 5546-H, 5547-C
and SCHAIBLE 957

FS5-7C **FS5-7H**
Fits BRIGGS
OEM 22097 RH-C,
22098 LH-H:
T-8802-03 Slant Back

FS5-8C **FS5-8H**
Fits BRIGGS
OEM 22033-C RH,
22034-H LH:
T-8805-15
Shelf Back Lav.

FS5-11C **FS5-11H**
Fits KOHLER
OEM 32491: K-8178
Single Lav. Fitting

FS5-14C **FS5-14H**
Fits SPEAKMAN
OEM G3-140-H,
G3-139-C: S-4700

FS5-15C **FS5-15H**
Fits STERLING
OEM 99S-3193-H,
99S-3194-C

FS5-16C **FS5-16H**
Fits CRANE Sleeve Unit
(25A-10-3/4),
C-31806 Telsa,
C-32267 Securo Jr.

FS5-18C **FS5-18H**
Fits SPEAKMAN
OEM 3-250-H, 3-251-C:
S-4095 Diamond Lav.
Fitting
Also fits RICHMOND

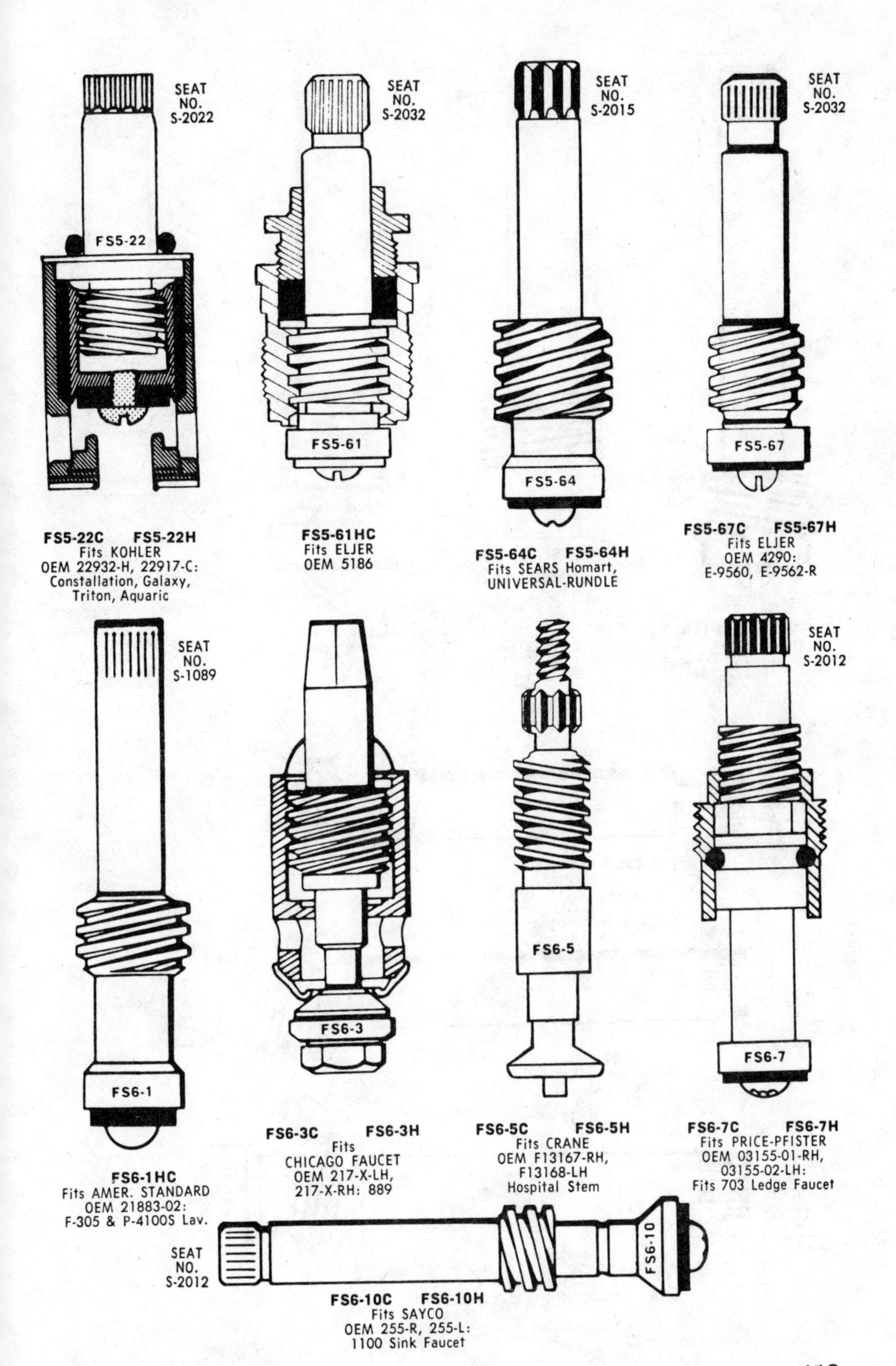
SEAT
NO.
S-2022
FS5-22
SEAT
NO.
S-2032
FS5-61
SEAT
NO.
S-2015
FS5-64
SEAT
NO.
S-2032
FS5-67
FS5-22C FS5-22H
Fits KOHLER
OEM 22932-H, 22917-C:
Constallation, Galaxy,
Triton, Aquaric
FS5-61HC
Fits ELJER
OEM 5186
FS5-64C FS5-64H
Fits SEARS Homart,
UNIVERSAL-RUNDLE
FS5-67C FS5-67H
Fits ELJER
OEM 4290:
E-9560, E-9562-R
SEAT
NO.
S-1089
FS6-1
SEAT
NO.
S-2012
FS6-3
FS6-5
FS6-7
FS6-3C FS6-3H
Fits
CHICAGO FAUCET
OEM 217-X-LH,
217-X-RH: 889
FS6-5C FS6-5H
Fits CRANE
OEM F13167-RH,
F13168-LH
Hospital Stem
FS6-7C FS6-7H
Fits PRICE-PFISTER
OEM 03155-01-RH,
03155-02-LH:
Fits 703 Ledge Faucet
FS6-1HC
Fits AMER. STANDARD
OEM 21883-02:
F-305 & P-4100S Lav.
SEAT
NO.
S-2012
FS6-10
FS6-10C FS6-10H
Fits SAYCO
OEM 255-R, 255-L:
1100 Sink Faucet

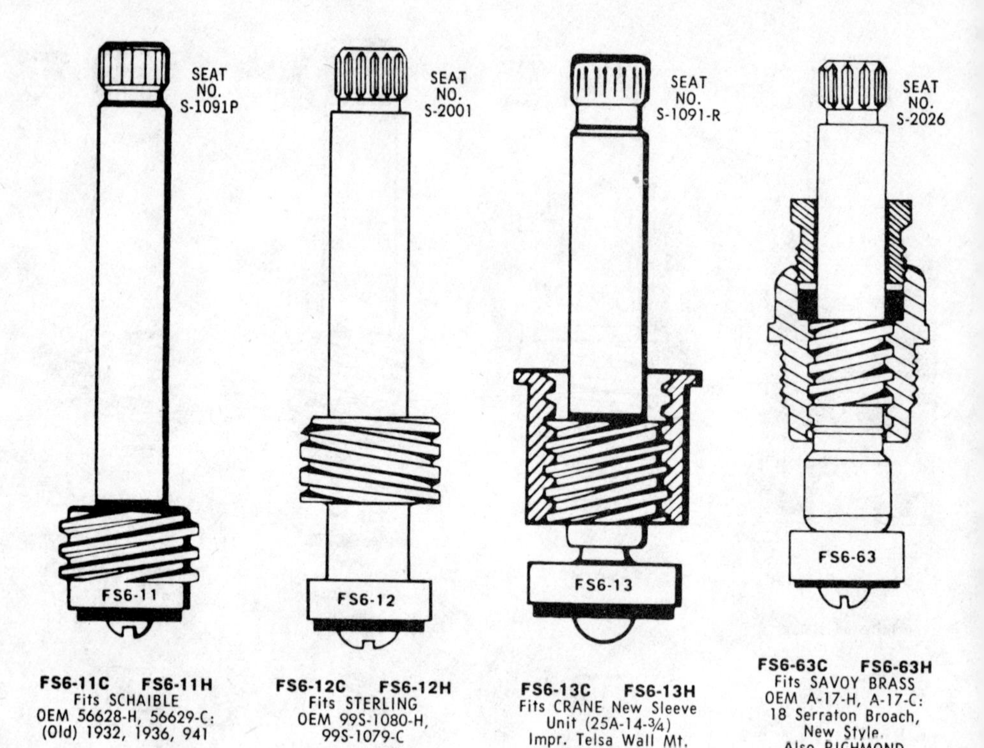

FS6-11C FS6-11H
Fits SCHAIBLE
OEM 56628-H, 56629-C:
(Old) 1932, 1936, 941

FS6-12C FS6-12H
Fits STERLING
OEM 99S-1080-H,
99S-1079-C

FS6-13C FS6-13H
Fits CRANE New Sleeve
Unit (25A-14-¾)
Impr. Telsa Wall Mt.
Sink Faucet

FS6-63C FS6-63H
Fits SAVOY BRASS
OEM A-17-H, A-17-C:
18 Serraton Broach,
New Style.
Also RICHMOND

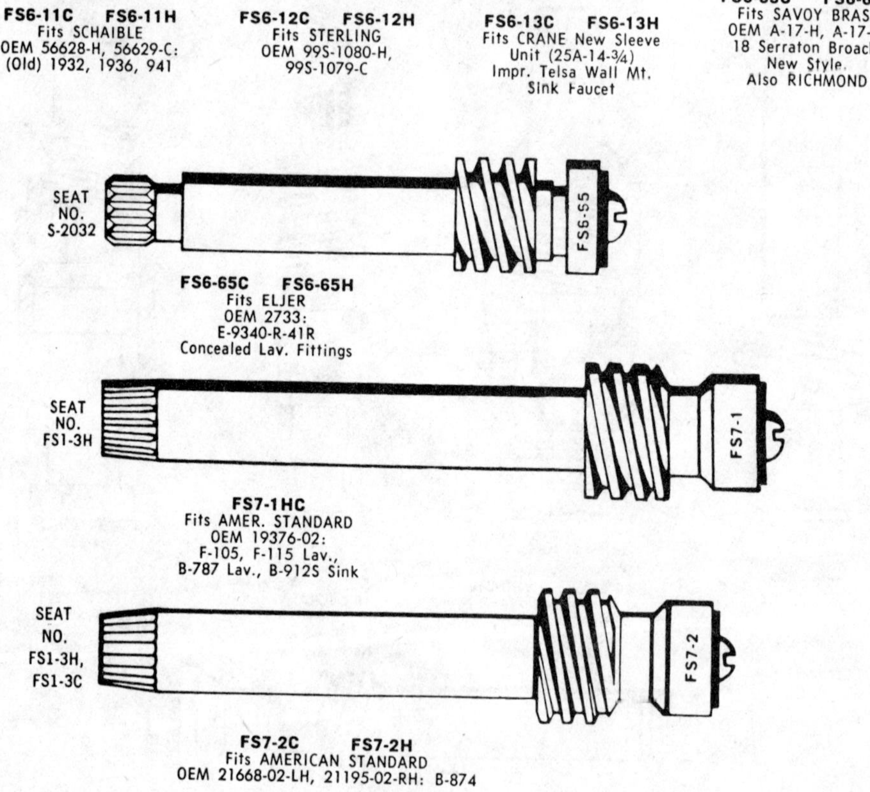

FS6-65C FS6-65H
Fits ELJER
OEM 2733:
E-9340-R-41R
Concealed Lav. Fittings

FS7-1HC
Fits AMER. STANDARD
OEM 19376-02:
F-105, F-115 Lav.,
B-787 Lav., B-912S Sink

FS7-2C FS7-2H
Fits AMERICAN STANDARD
OEM 21668-02-LH, 21195-02-RH: B-874

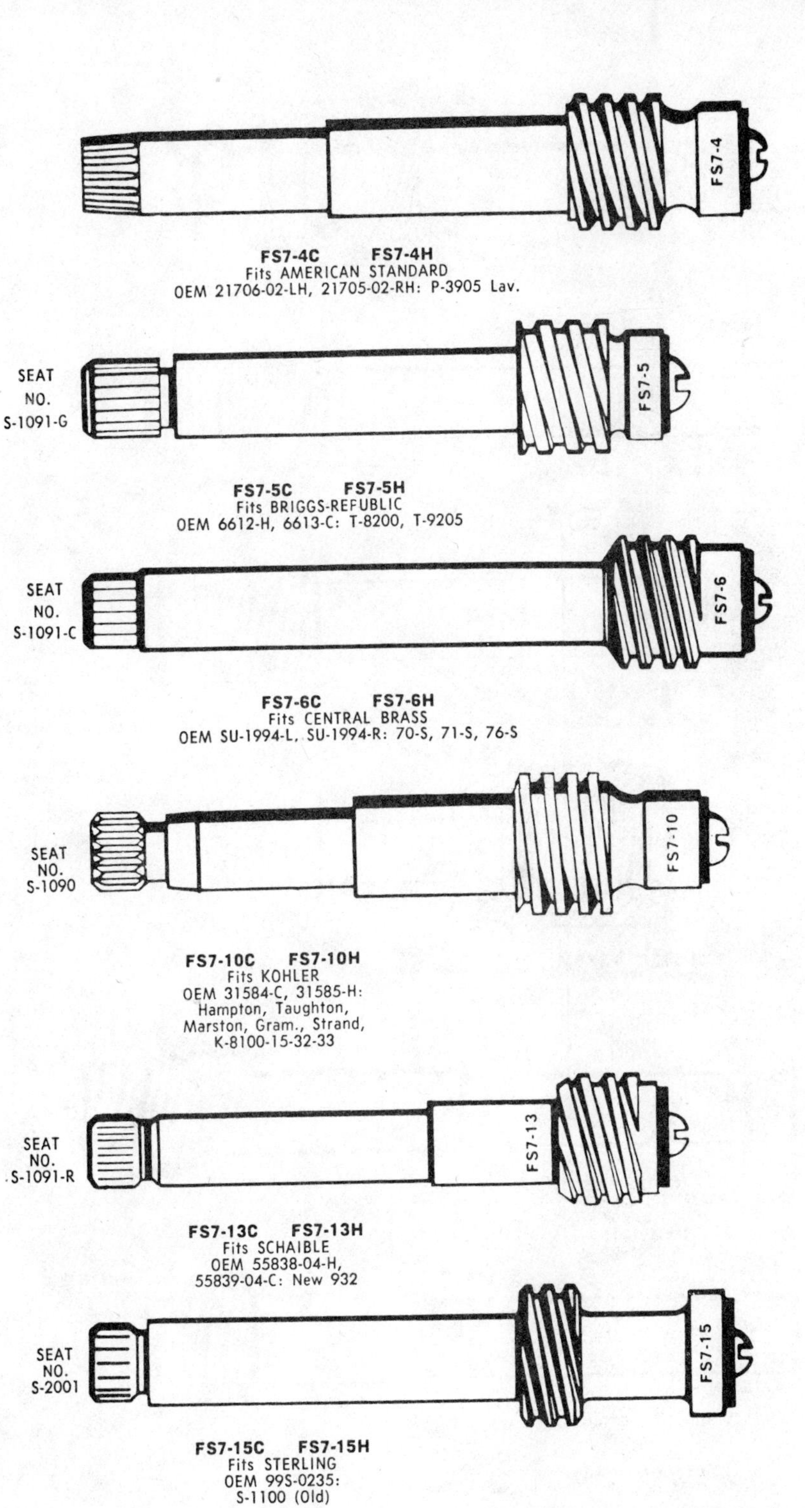

FS7-4C **FS7-4H**
Fits AMERICAN STANDARD
OEM 21706-02-LH, 21705-02-RH: P-3905 Lav.

FS7-5C **FS7-5H**
Fits BRIGGS-REFUBLIC
OEM 6612-H, 6613-C: T-8200, T-9205

FS7-6C **FS7-6H**
Fits CENTRAL BRASS
OEM SU-1994-L, SU-1994-R: 70-S, 71-S, 76-S

FS7-10C **FS7-10H**
Fits KOHLER
OEM 31584-C, 31585-H:
Hampton, Taughton,
Marston, Gram., Strand,
K-8100-15-32-33

FS7-13C **FS7-13H**
Fits SCHAIBLE
OEM 55838-04-H,
55839-04-C: New 932

FS7-15C **FS7-15H**
Fits STERLING
OEM 99S-0235:
S-1100 (Old)

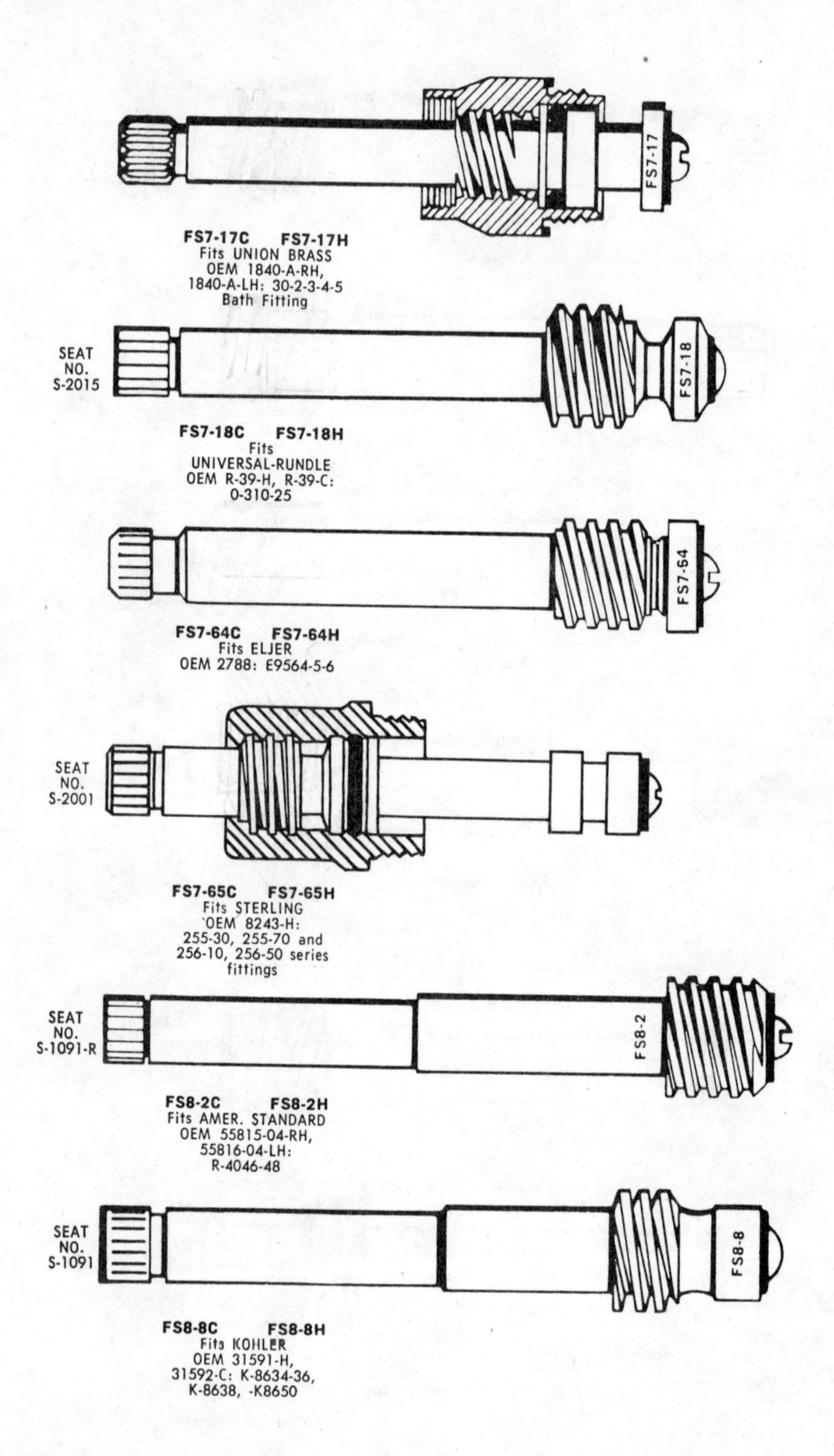
FS7-17
FS7-17C FS7-17H
Fits UNION BRASS
OEM 1840-A-RH,
1840-A-LH: 30-2-3-4-5
Bath Fitting
SEAT
NO.
S-2015
FS7-18
FS7-18C FS7-18H
Fits
UNIVERSAL-RUNDLE
OEM R-39-H, R-39-C:
0-310-25
FS7-64
FS7-64C FS7-64H
Fits ELJER
OEM 2788: E9564-5-6
SEAT
NO.
S-2001
FS7-65C FS7-65H
Fits STERLING
OEM 8243-H:
255-30, 255-70 and
256-10, 256-50 series
fittings
SEAT
NO.
S-1091-R
FS8-2
FS8-2C FS8-2H
Fits AMER. STANDARD
OEM 55815-04-RH,
55816-04-LH:
R-4046-48
SEAT
NO.
S-1091
FS8-8
FS8-8C FS8-8H
Fits KOHLER
OEM 31591-H,
31592-C: K-8634-36,
K-8638, -K8650

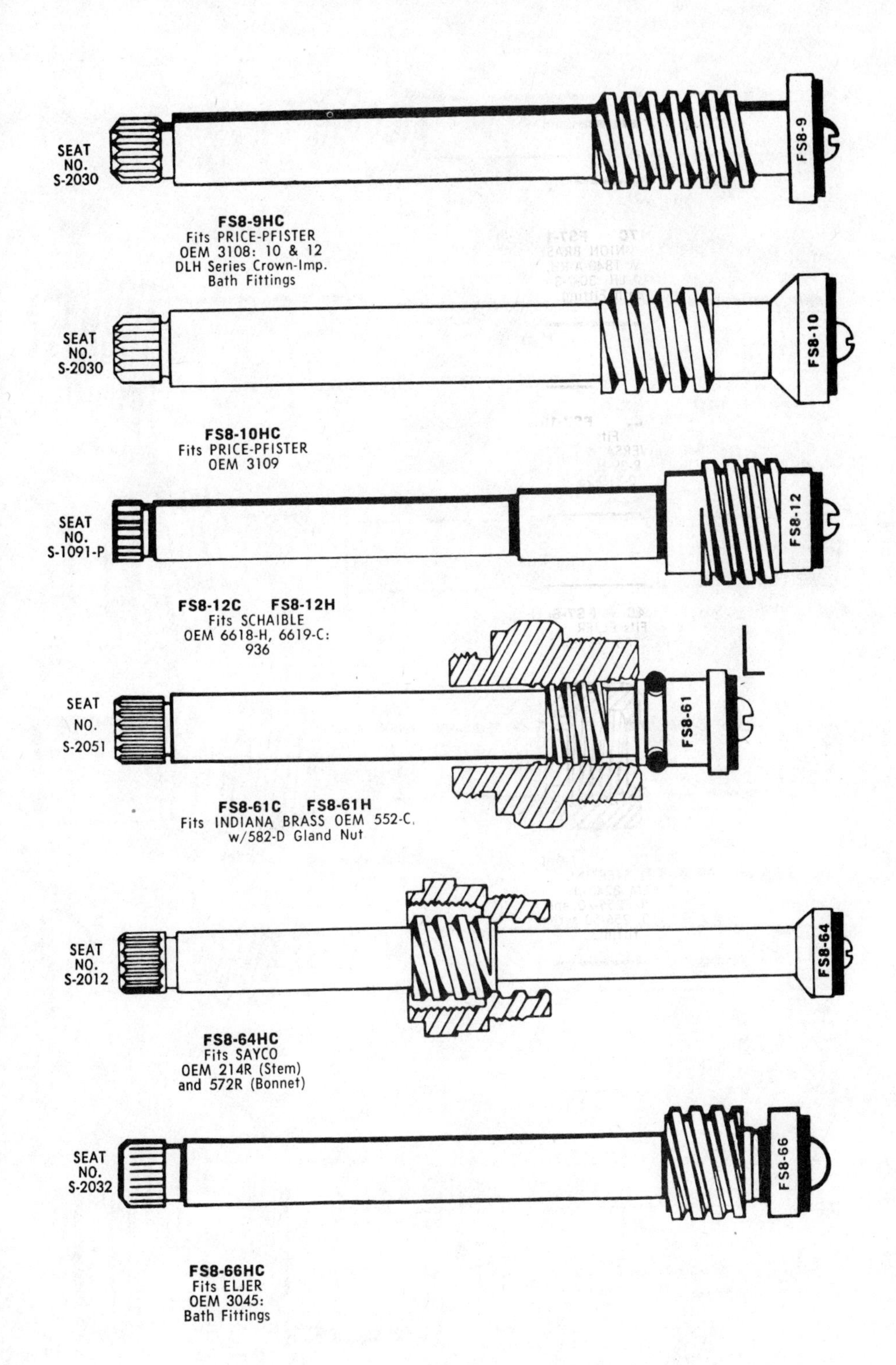
SEAT
NO.
S-2030
FS8-9
FS8-9HC
Fits PRICE-PFISTER
OEM 3108: 10 & 12
DLH Series Crown-Imp.
Bath Fittings
SEAT
NO.
S-2030
FS8-10
FS8-10HC
Fits PRICE-PFISTER
OEM 3109
SEAT
NO.
S-1091-P
FS8-12
FS8-12C FS8-12H
Fits SCHAIBLE
OEM 6618-H, 6619-C:
936
SEAT
NO.
S-2051
FS8-61
FS8-61C FS8-61H
Fits INDIANA BRASS OEM 552-C,
w/582-D Gland Nut
SEAT
NO.
S-2012
FS8-64
FS8-64HC
Fits SAYCO
OEM 214R (Stem)
and 572R (Bonnet)
SEAT
NO.
S-2032
FS8-66
FS8-66HC
Fits ELJER
OEM 3045:
Bath Fittings

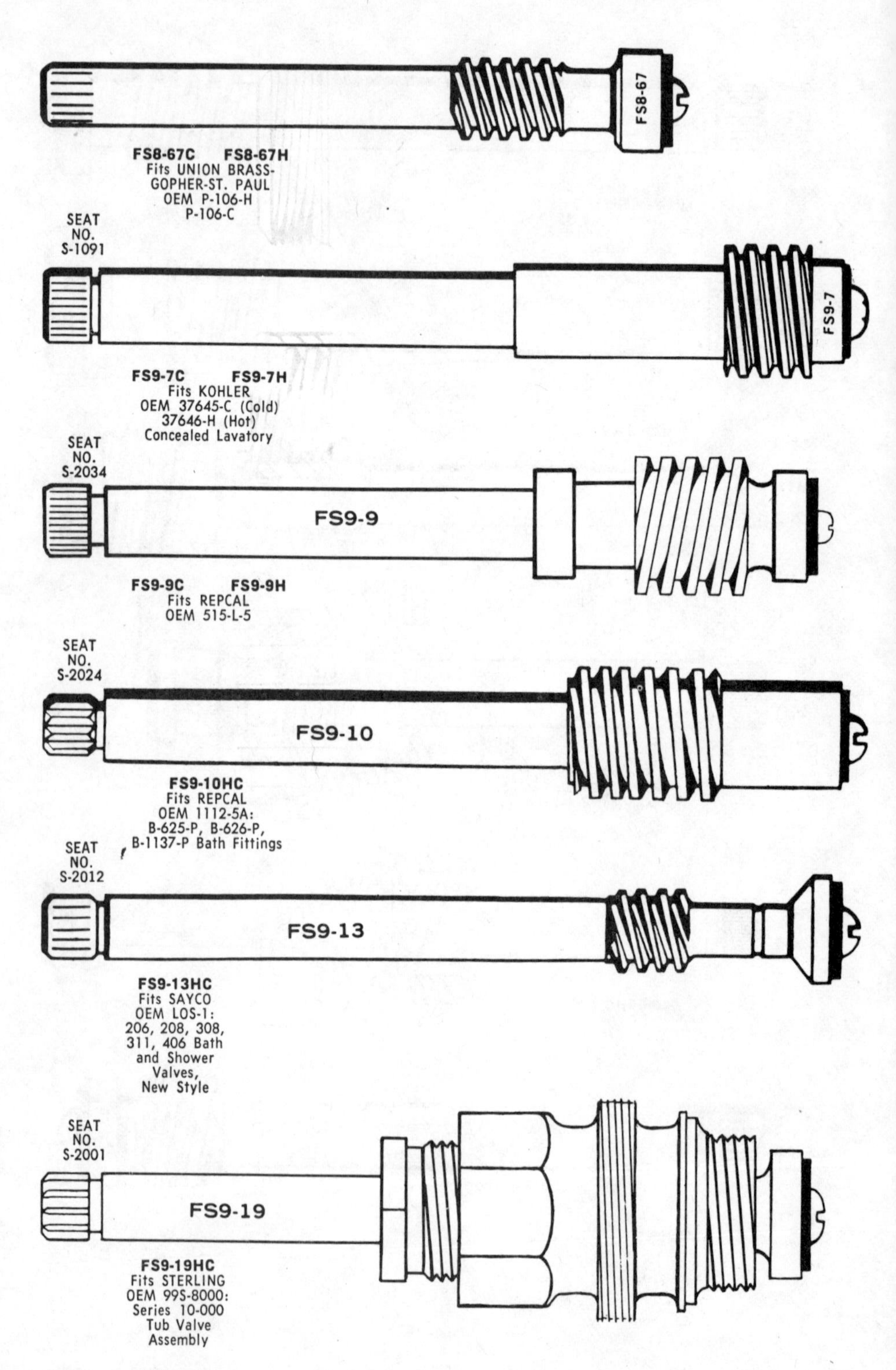
FS8-67
FS8-67C FS8-67H
Fits UNION BRASS-
GOPHER-ST. PAUL
OEM P-106-H
P-106-C
SEAT
NO.
S-1091
FS9-7
FS9-7C FS9-7H
Fits KOHLER
OEM 37645-C (Cold)
37646-H (Hot)
Concealed Lavatory
SEAT
NO.
S-2034
FS9-9
FS9-9C FS9-9H
Fits REPCAL
OEM 515-L-5
SEAT
NO.
S-2024
FS9-10
FS9-10HC
Fits REPCAL
OEM 1112-5A:
B-625-P, B-626-P,
B-1137-P Bath Fittings
SEAT
NO.
S-2012
FS9-13
FS9-13HC
Fits SAYCO
OEM LOS-1:
206, 208, 308,
311, 406 Bath
and Shower
Valves,
New Style
SEAT
NO.
S-2001
FS9-19
FS9-19HC
Fits STERLING
OEM 99S-8000:
Series 10-000
Tub Valve
Assembly

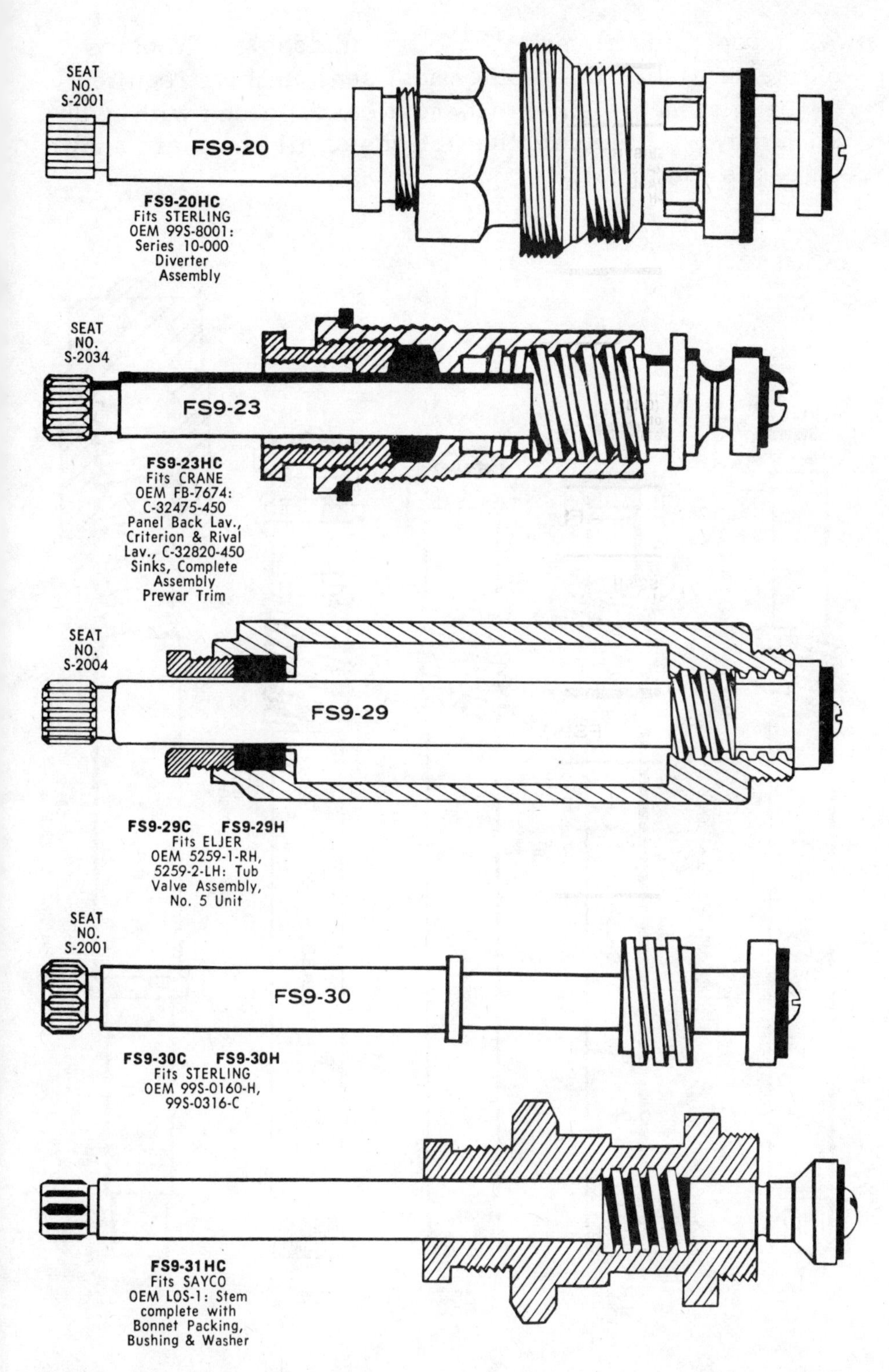
SEAT
NO.
S-2001
FS9-20
FS9-20HC
Fits STERLING
OEM 99S-8001:
Series 10-000
Diverter
Assembly
SEAT
NO.
S-2034
FS9-23
FS9-23HC
Fits CRANE
OEM FB-7674:
C-32475-450
Panel Back Lav.,
Criterion & Rival
Lav., C-32820-450
Sinks, Complete
Assembly
Prewar Trim
SEAT
NO.
S-2004
FS9-29
FS9-29C **FS9-29H**
Fits ELJER
OEM 5259-1-RH,
5259-2-LH: Tub
Valve Assembly,
No. 5 Unit
SEAT
NO.
S-2001
FS9-30
FS9-30C **FS9-30H**
Fits STERLING
OEM 99S-0160-H,
99S-0316-C
FS9-31HC
Fits SAYCO
OEM LOS-1: Stem
complete with
Bonnet Packing,
Bushing & Washer

When replacing stems, it is often advisable to replace old seats also. The replacement seat numbers required for each stem are shown (in most cases) along with each stem drawing. Actual illustrations of all seats are shown on pages 128 to 131.

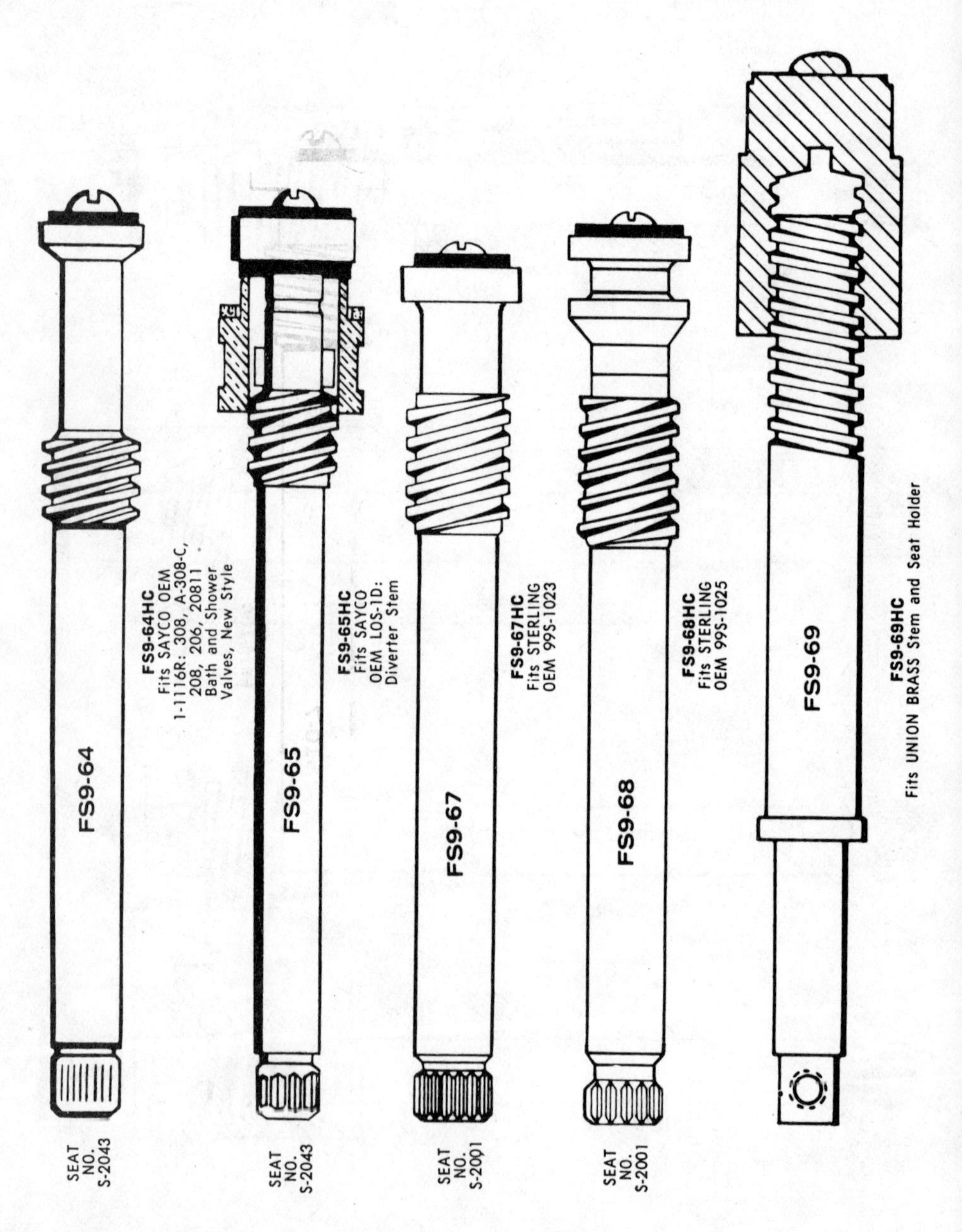

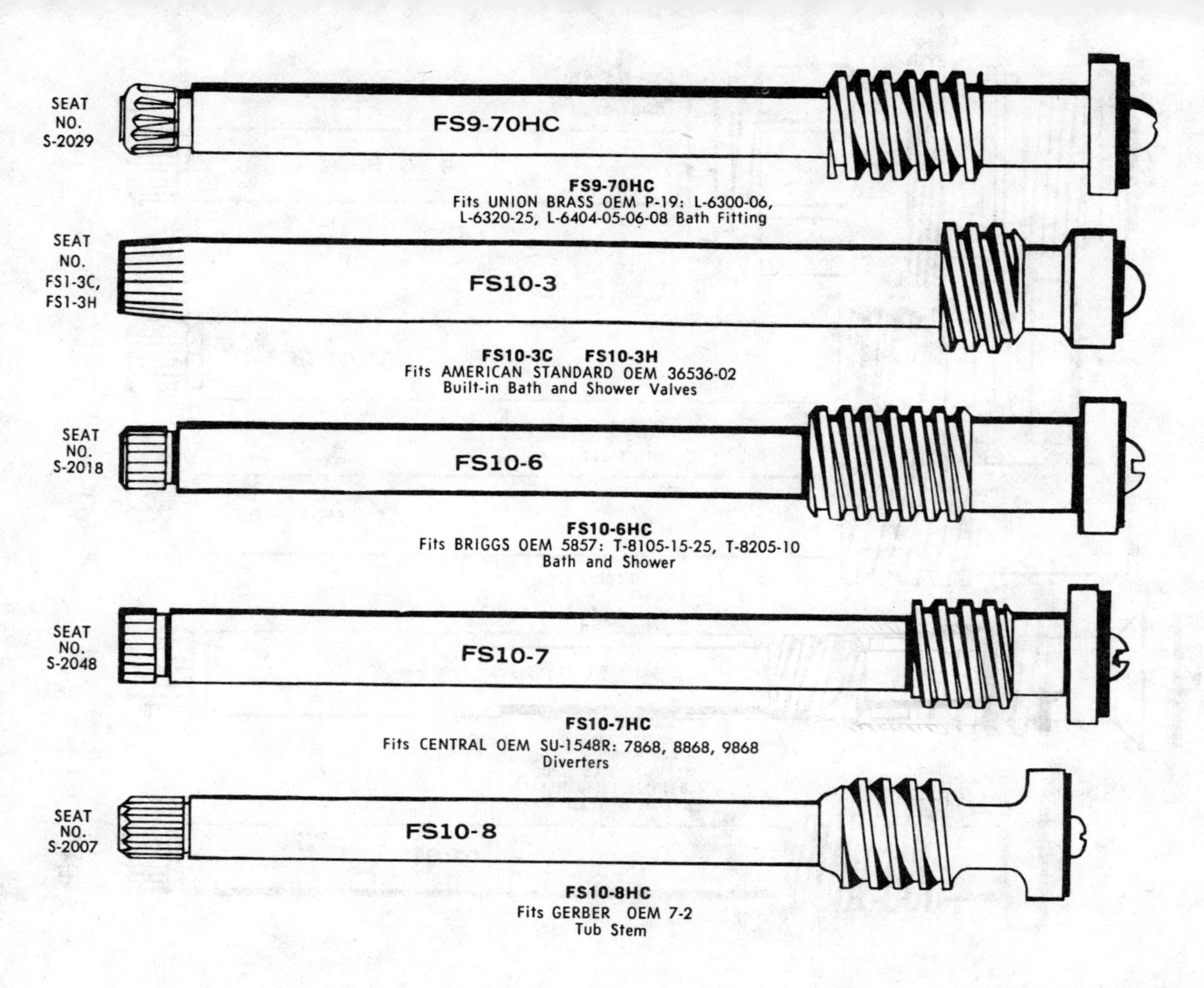

FS9-70HC
Fits UNION BRASS OEM P-19: L-6300-06, L-6320-25, L-6404-05-06-08 Bath Fitting

FS10-3C FS10-3H
Fits AMERICAN STANDARD OEM 36536-02 Built-in Bath and Shower Valves

FS10-6HC
Fits BRIGGS OEM 5857: T-8105-15-25, T-8205-10 Bath and Shower

FS10-7HC
Fits CENTRAL OEM SU-1548R: 7868, 8868, 9868 Diverters

FS10-8HC
Fits GERBER OEM 7-2 Tub Stem

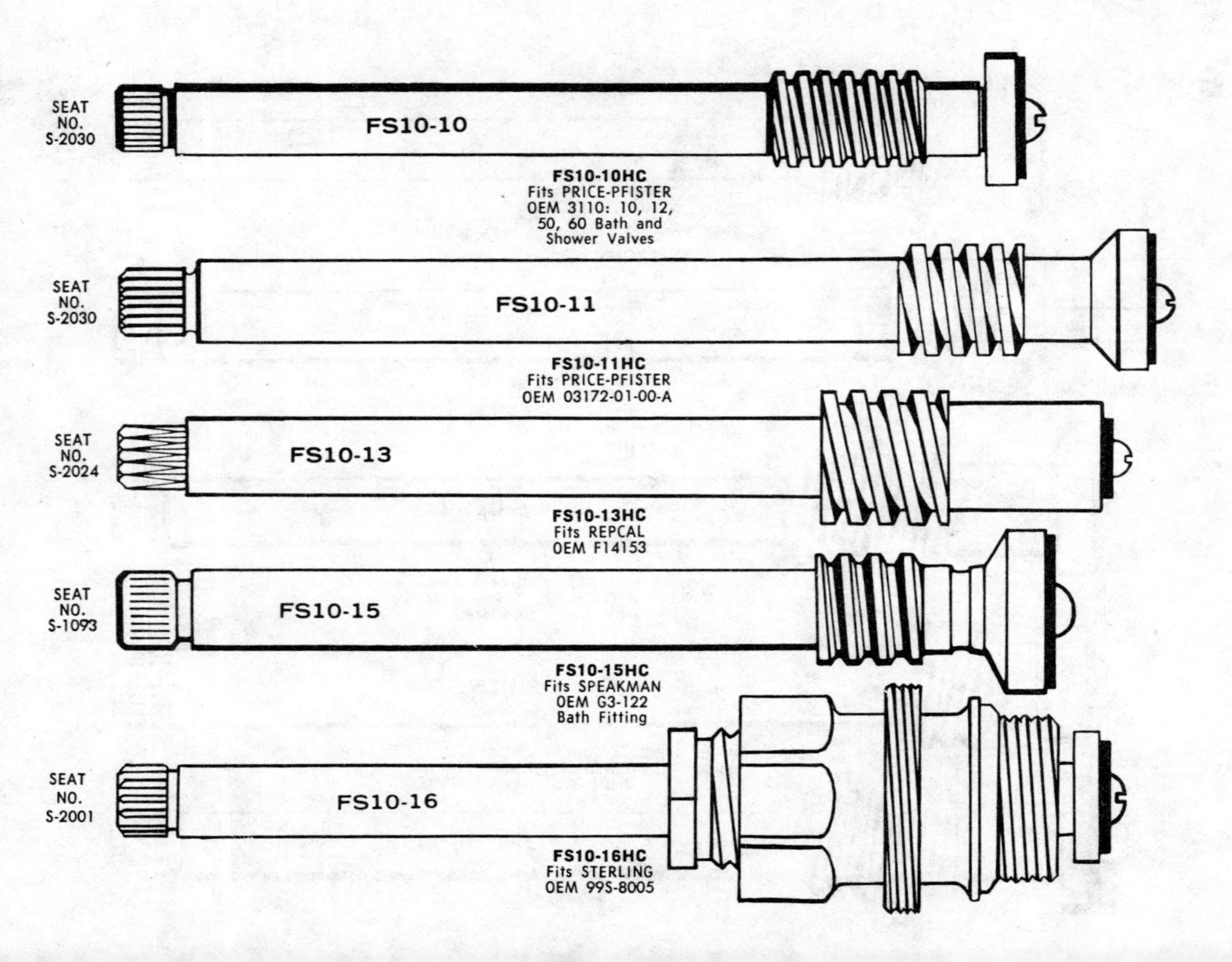
SEAT NO. S-2030
FS10-10
FS10-10HC
Fits PRICE-PFISTER
OEM 3110: 10, 12, 50, 60 Bath and Shower Valves
SEAT NO. S-2030
FS10-11
FS10-11HC
Fits PRICE-PFISTER
OEM 03172-01-00-A
SEAT NO. S-2024
FS10-13
FS10-13HC
Fits REPCAL
OEM F14153
SEAT NO. S-1093
FS10-15
FS10-15HC
Fits SPEAKMAN
OEM G3-122
Bath Fitting
SEAT NO. S-2001
FS10-16
FS10-16HC
Fits STERLING
OEM 99S-8005

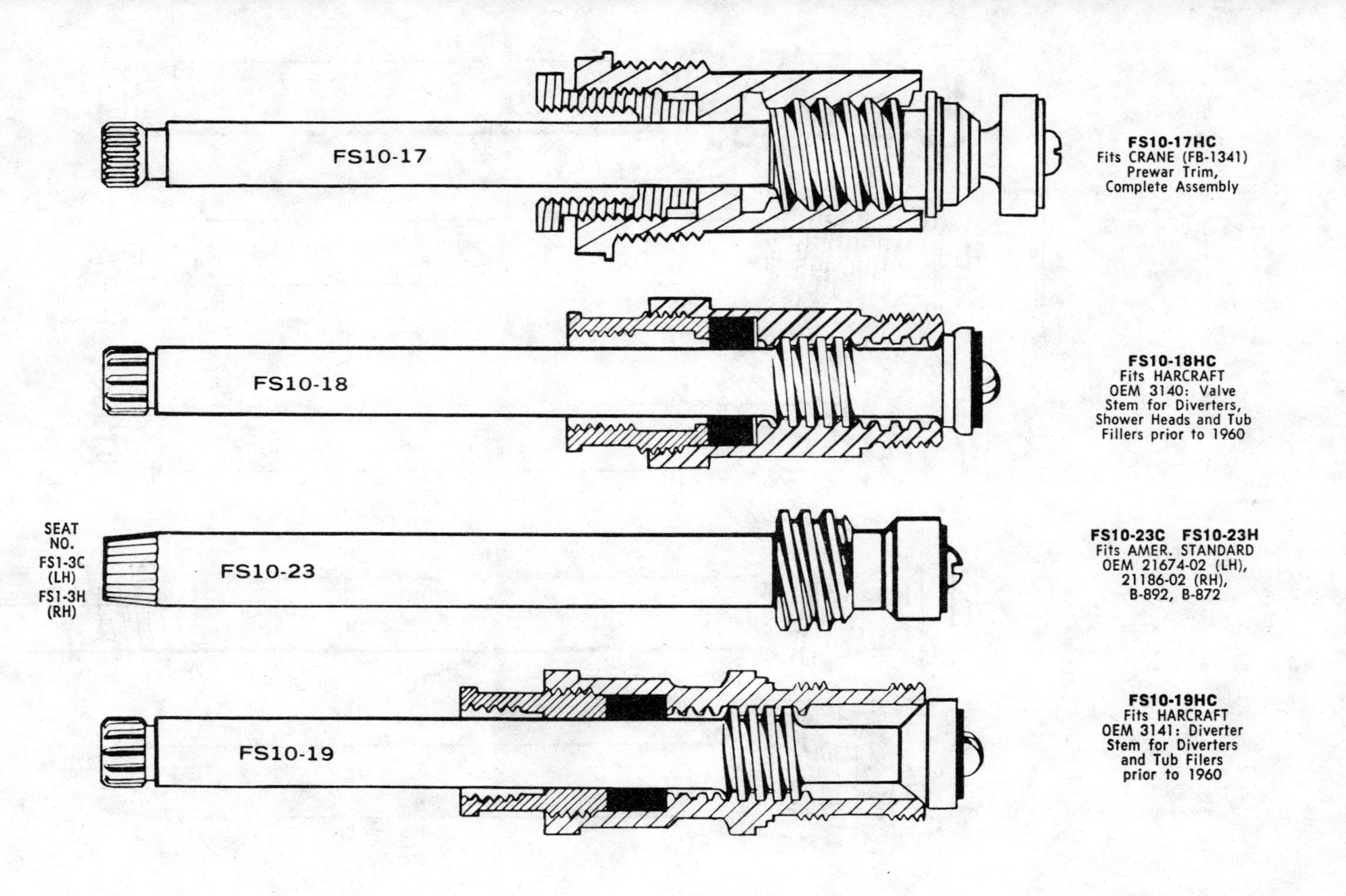

FS10-17HC
Fits CRANE (FB-1341)
Prewar Trim,
Complete Assembly

FS10-18HC
Fits HARCRAFT
OEM 3140: Valve
Stem for Diverters,
Shower Heads and Tub
Fillers prior to 1960

FS10-23C FS10-23H
Fits AMER. STANDARD
OEM 21674-02 (LH),
21186-02 (RH),
B-892, B-872

FS10-19HC
Fits HARCRAFT
OEM 3141: Diverter
Stem for Diverters
and Tub Filers
prior to 1960

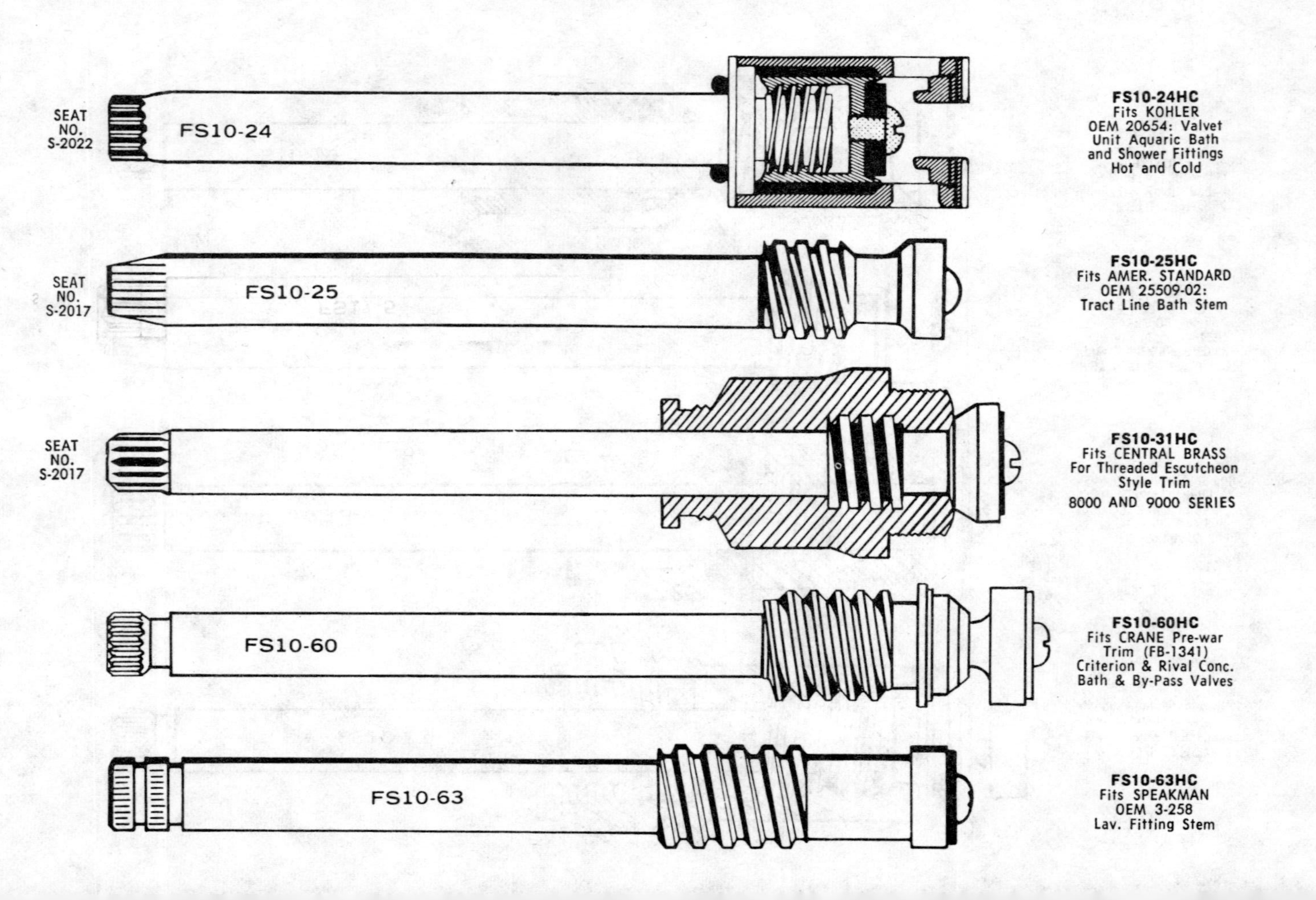
SEAT
NO.
S-2022
FS10-24
FS10-24HC
Fits KOHLER
OEM 20654: Valvet
Unit Aquaric Bath
and Shower Fittings
Hot and Cold
SEAT
NO.
S-2017
FS10-25
FS10-25HC
Fits AMER. STANDARD
OEM 25509-02:
Tract Line Bath Stem
SEAT
NO.
S-2017
FS10-31HC
Fits CENTRAL BRASS
For Threaded Escutcheon
Style Trim
8000 AND 9000 SERIES
FS10-60
FS10-60HC
Fits CRANE Pre-war
Trim (FB-1341)
Criterion & Rival Conc.
Bath & By-Pass Valves
FS10-63
FS10-63HC
Fits SPEAKMAN
OEM 3-258
Lav. Fitting Stem

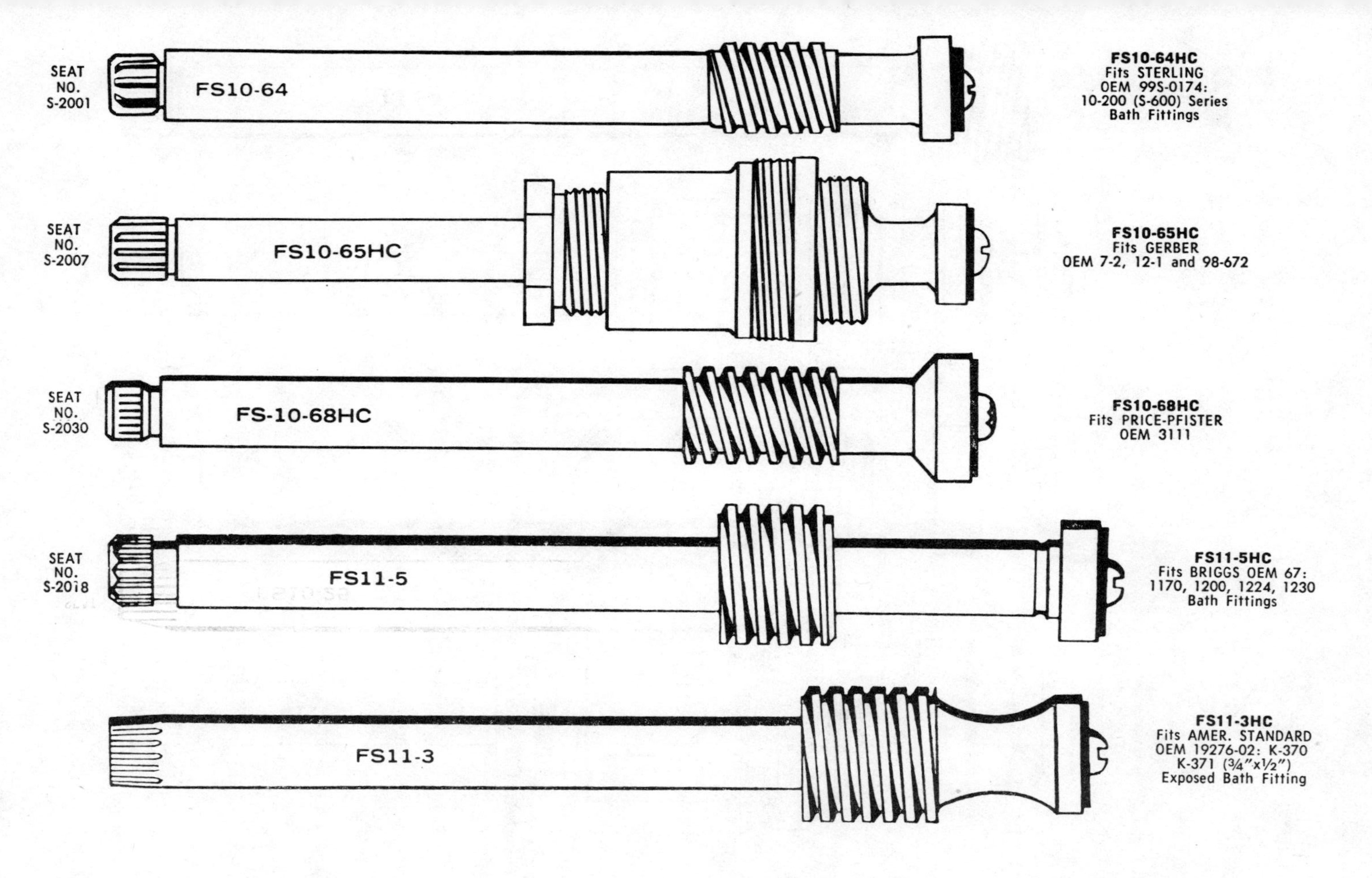
SEAT
NO.
S-2001
FS10-64
FS10-64HC
Fits STERLING
OEM 99S-0174:
10-200 (S-600) Series
Bath Fittings
SEAT
NO.
S-2007
FS10-65HC
FS10-65HC
Fits GERBER
OEM 7-2, 12-1 and 98-672
SEAT
NO.
S-2030
FS-10-68HC
FS10-68HC
Fits PRICE-PFISTER
OEM 3111
SEAT
NO.
S-2018
FS11-5
FS11-5HC
Fits BRIGGS OEM 67:
1170, 1200, 1224, 1230
Bath Fittings
FS11-3
FS11-3HC
Fits AMER. STANDARD
OEM 19276-02: K-370
K-371 (3/4"x1/2")
Exposed Bath Fitting

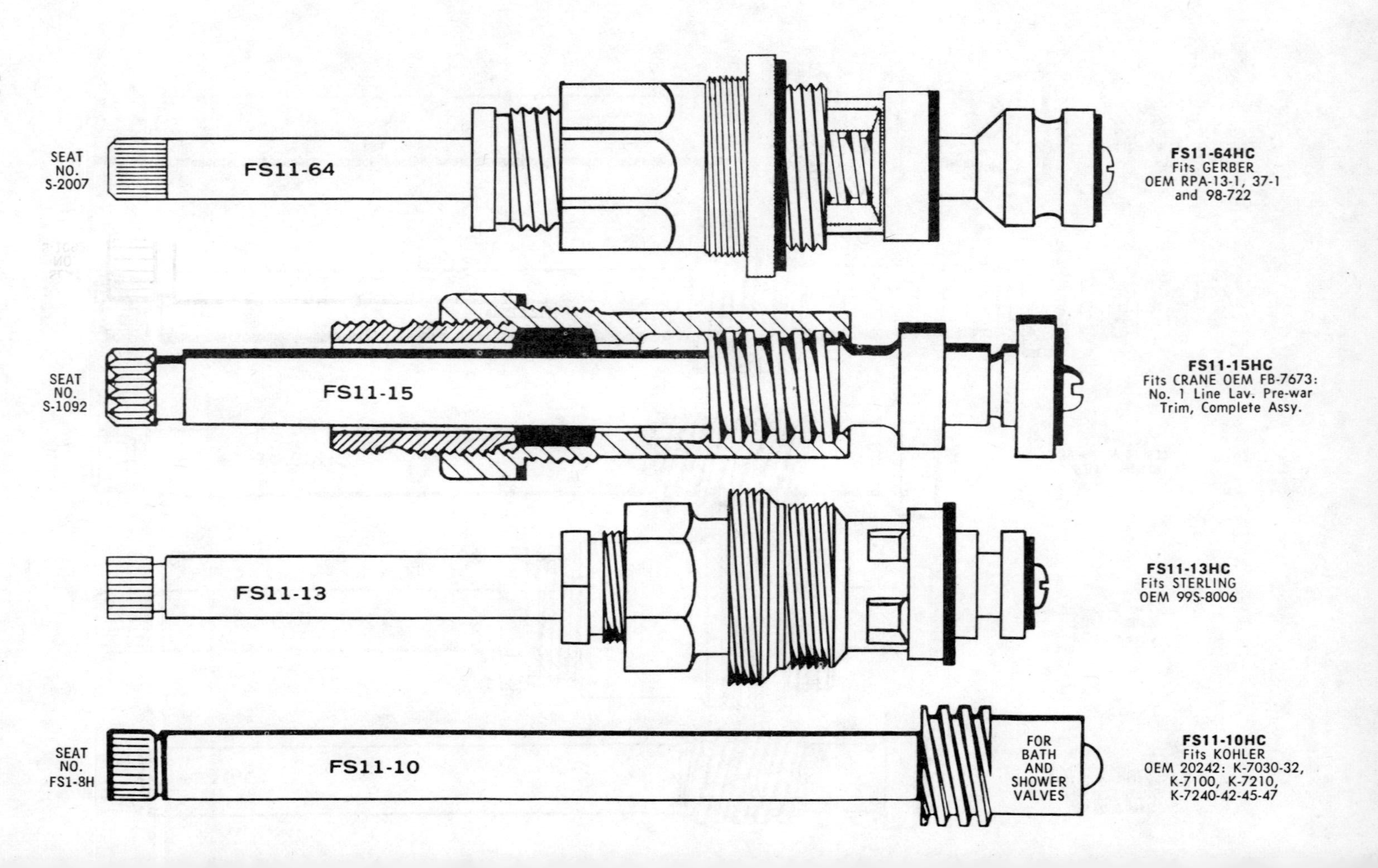
SEAT
NO.
S-2007
FS11-64
FS11-64HC
Fits GERBER
OEM RPA-13-1, 37-1
and 98-722
SEAT
NO.
S-1092
FS11-15
FS11-15HC
Fits CRANE OEM FB-7673:
No. 1 Line Lav. Pre-war
Trim, Complete Assy.
FS11-13
FS11-13HC
Fits STERLING
OEM 99S-8006
SEAT
NO.
FS1-8H
FS11-10
FOR
BATH
AND
SHOWER
VALVES
FS11-10HC
Fits KOHLER
OEM 20242: K-7030-32,
K-7100, K-7210,
K-7240-42-45-47

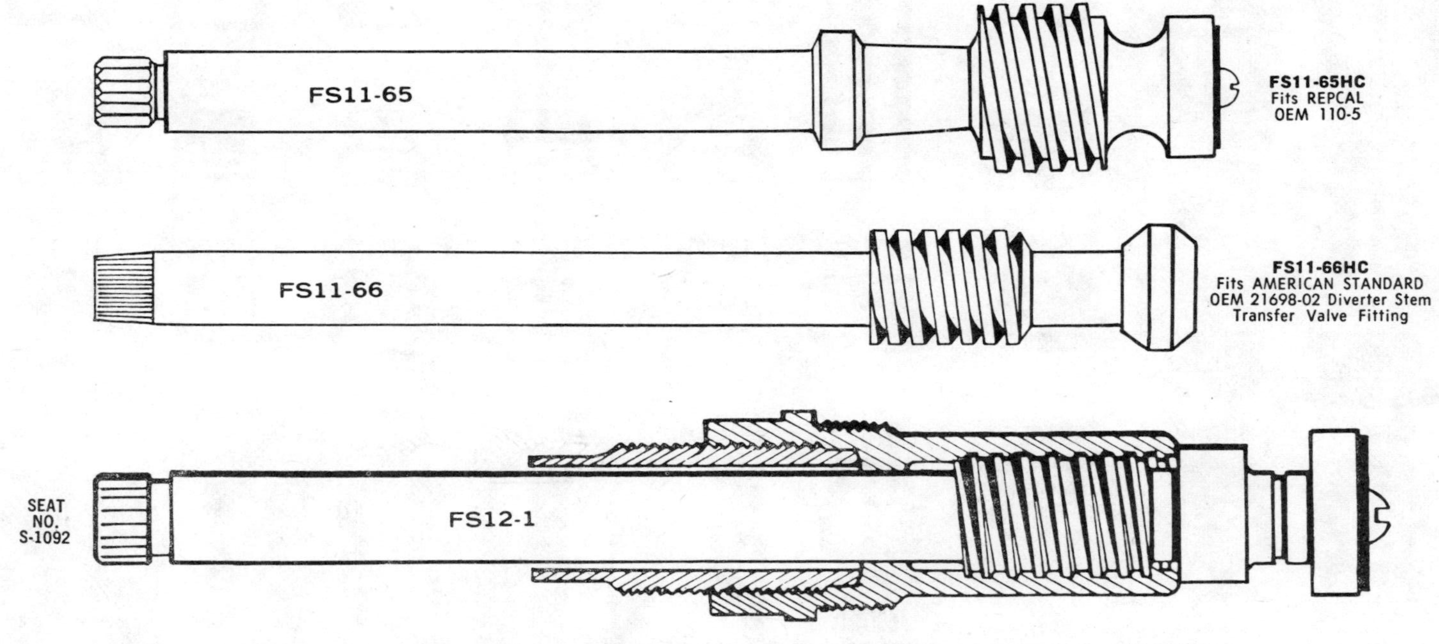

FS12-1HC Fits CRANE OEM FB-1077: Complete Assembly, Pre-war Trim, No. 1 Line Bath Valves, ½" and ¾"

RENEW SEATS (ACTUAL SIZE ILLUSTRATIONS)

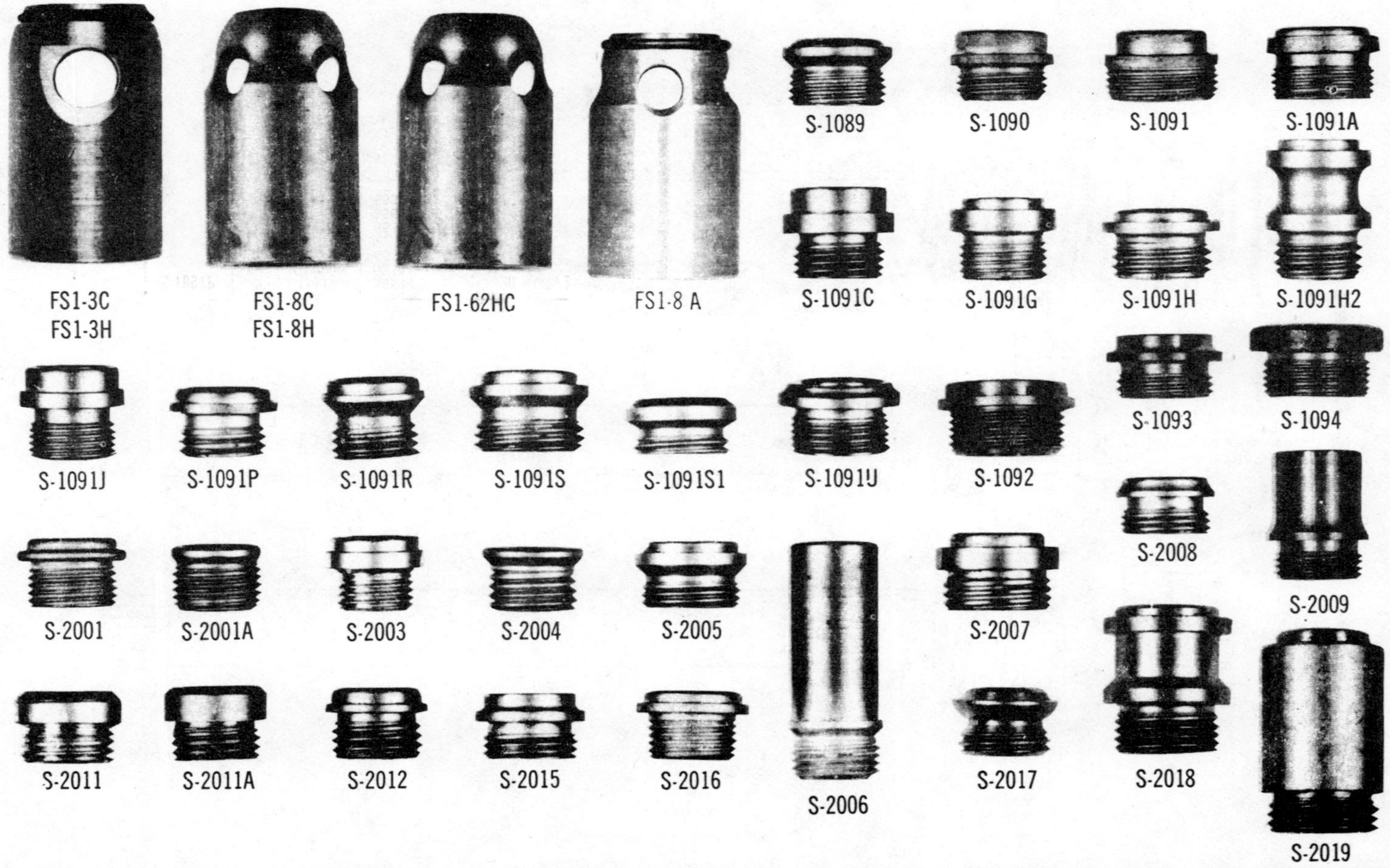

FS1-3C
FS1-3H

FS1-8C
FS1-8H

FS1-62HC

FS1-8 A

S-1089

S-1090

S-1091

S-1091A

S-1091C

S-1091G

S-1091H

S-1091H2

S-1093

S-1094

S-1091J

S-1091P

S-1091R

S-1091S

S-1091S1

S-1091U

S-1092

S-2008

S-2009

S-2001

S-2001A

S-2003

S-2004

S-2005

S-2007

S-2011

S-2011A

S-2012

S-2015

S-2016

S-2006

S-2017

S-2018

S-2019

Part No.	Manufacturer of Faucet	O.E.M. Part No.	Specifications Thd. - O.D. (Thd.) - Ht.
FS1-3C	**Amer. Standard**	**20563-08**	Barrel Seat Cold Side (L.H. Thread)w/O-Rings
FS1-3H	**Amer. Standard**	**20336-08**	Barrel Seat Hot Side (R.H. Thread)w/O-Rings
FS1-8C	**Kohler**	**39717**	Barrel Seat Cold Side (L.H. Thread)
FS1-8H	**Kohler**	**32462**	Barrel Seat Hot Side (R.H. Thread)
FS1-62HC	**Kohler**	**32473**	Barrel Seat (Not Threaded)
S-1089	**Amer. Standard**	**174-14**	**24 - 35/64 - 3/8**
S-1090	**Kohler**	**40602**	**27 - 1/2 - 25/64**
S-1091	**Kohler**	**33345**	**27 - 5/8 - 25/64**
S-1091A	**Dick Brothers**	**3055/4010**	**27 - 9/16 - 7/16**
S-1091C	**Central Brass**	**263-B**	**24 - 1/2 - 15/32**
S-1091G	**Briggs-Republic**	**22092**	**24 - 1/2 - 15/32**
S-1091H	**Briggs-Republic**	**8777**	**20 - 9/16 - 3/8**
S-1091H2	**Briggs-Republic**	**22040**	**20 - 1/2 - 3/4**
S-1091J	**Eljer**	**2734**	**27 - 1/2 - 1/2**
S-1091P	Amer. Standard Amer. Kitchens Crosley Eljer Schaible Tracy Youngstown	56534-07 6534 6534 .... 6534 8662-8663 6534	**20 - 1/2 - 3/8**
S-1091R	Amer. Standard Schaible Youngstown	57281-07 5506 5506	**20 - 1/2 - 7/16**
S-1091S	**Tracy**	**8435**	**20 - 5/8 - 7/16**
S-1091S1	Amer. Standard Schaible Sears Youngstown	**12002-07**	**20 - 35/64 - 5/16**
S-1091U	**Speakman**	**05-0305** Fmrly. S-5460	**28 - 9/16 - 3/8**
S-1092	**Crane**	**F-5914**	**22 - 43/64 - 27/64**
S-1093	**Burlington Br.**	**6-1**	**24 - 17/32 - 23/64**
S-1094	**Speakman**	**5-451**	**24 - 5/8 - 7/16**
S-2001	Amer. Kitchen Sterling United Valve	802464 99S-0369	**24 - 1/2 - 3/8**
S-2001A	**Empire Brass**	**20**	**24 - 1/2 - 11/32**
S-2003	American Brass Amer. Standard	 862-14	**24 - 7/16 - 13/32**
S-2004	Barnes Eljer	7501 5257	**20 - 1/2 - 11/32**
S-2005	**Barnes**	**7302**	**24 - 9/16 - 3/8**
S-2006	**Empire Brass**	**510**	**24 - 1/2 - 1¼**
S-2007	**Gerber**	**16**	**20 - 5/8 - 7/16**
S-2008	Indiana Brass Milwaukee	638-0 964	**27 - 1/2 - 3/8**
S-2009	**Milwaukee**	**3169**	**27 - 1/2 - 23/32**
S-2011	Crane Savoy Sayco	.... 33 6A3	**20 - 35/64 - 7/16**
S-2011A	**Savoy**	**524**	**20 - 1/2 - 13/32**
S-2012	Price-Pfister Sayco	05716-01 6	**20 - 1/2 - 3/8**
S-2015	**Univ.-Rundle**	**12R**	**24 - 17/32 - 3/8**
S-2016	**Wolverine**	**586S**	**24 - 1/2 - 3/8**
S-2017	**Amer. Standard**	**1848-07**	**18 - 17/32 - 3/8**
S-2018	**Briggs-Republic**	**91**	**20 - 5/8 - 27/32**
S-2019	**Gerber**	**98**	**20 - 5/8 - 1⅛**

RENEW SEATS (ACTUAL SIZE ILLUSTRATIONS)

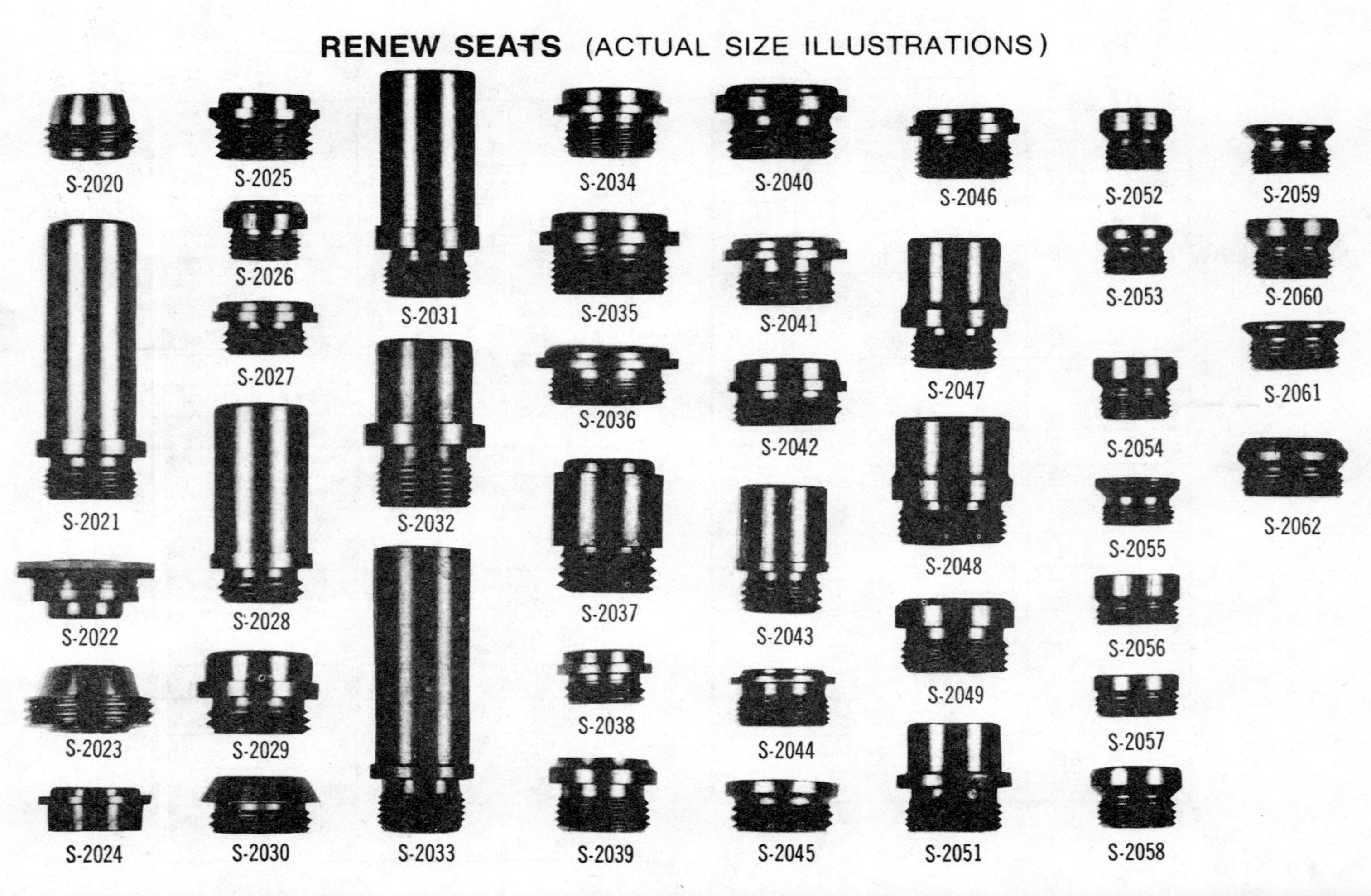

Part No.	Manufacturer of Faucet	O.E.M. Part No.	Specifications Thd. - O.D. (Thd.) - Ht.
S-2020	**Harcraft**	**6311**	**20 - 9/16 - 3/8**
S-2021	**Indiana**	**550-0**	**27 - 9/16 - 1⅝**
S-2022	**Kohler**	**23004**	No Thd. **- 25/32 - 21/64**
S-2023	**Harcraft**	**5377**	**20 - 3/4 - 3/8**
S-2024	**Repcal**	**1112-18A**	No Thd. **- 5/8 - 1/4**
S-2025	**Repcal**	F14301 1112-18B	**18 - 9/16 - 3/8**
S-2026 S-2010)	Speakman Richmond	05-0775 Fmrly. S-5465 K-12	**27 - 7/16 - 11/32**
S-2027 (S-2013)	Crane Union Brass	 2601	**18 - 1/2 - 5/16**
S-2028	**Union Brass**	**2606**	**18 - 1/2 - 1 5/32**
S-2029	**Union Brass**	**2602**	**18 - 5/8 - 15/32**
S-2030	**Price-Pfister**	**05713-01**	**18 - 21/32 - 3/8**
S-2031	**Eljer**	**2924**	**27 - 1/2 - 1 9/32**
S-2032	**Eljer**	**2750**	**27 - 35/64 - 15/16**
S-2033	**Eljer**	**4703**	**27 - 1/2 - 1 19/32**
S-2034	**Crane**	**F5913**	**22 - 9/16 - 13/32**
S-2035	**Crane**	**F5914**	**24 - 11/16 - 15/32**
S-2036	**Crane Speakman**		**24 - 11/16 - 11/32**
S-2037	Crane Michigan Brass	F3364 13117	**18 - 9/16 - 3/4**
S-2038	Crane Eljer Price-Pfister Sears	583 5257 5716-01 6A3	**24 - 1/2 - 5/16**
S-2039	**Briggs-Republic**	**6762**	**24 - 9/16 - 7/16**
S-2040	**Briggs-Republic**	**22379**	**20 - 5/8 - 7/16**
S-2041	**Wolverine**	**587-S**	**24 - 19/32 - 25/64**
S-2042	**Kohler**	**22526**	**27 - 5/8 - 3/8**
S-2043	**Sayco**	**6-468-C-39**	**20 - 1/2 - 3/4**
S-2044	**Repcal**	**1112-18**	**28 - 9/16 - 11/32**
S-2045	**Union Brass**	**2632**	**20 - 39/64 - 5/16**
S-2046	**Central**	**59**	**27 - 9/16 - 13/32**
S-2047	**Central**	**263**	**24 - 1/2 - 47/64**
S-2048	**Central**	**165**	**24 - 5/8 - 23/32**
S-2049	**Central**	**165/6500**	**24 - 39/64 - 27/64**
S-2051	**Indiana Brass**	**590-0**	**27 - 9/16 - 5/8**
S-2052	Sexauer or Skinner Tappg. Tools For use with Chicago, Economy,		**27 - 3/8 - 3/8**
S-2053	Sexauer or Skinner Tappg. Tools For use with Chicago, Economy,		**27 - 3/8 - 19/64**
S-2054	Sexauer or Skinner Tappg. Tools For use with Chicago, Economy,		**27 - 7/16 - 3/8**
S-2055	Sexauer or Skinner Tappg. Tools For use with Chicago, Economy,		**27 - 7/16 - 19/64**
S-2056	Sexauer or Skinner Tappg. Tools For use with Chicago, Economy,		**27 - 1/2 - 19/64**
S-2057	Sexauer or Skinner Tappg. Tools For use with Chicago, Economy,		**27 - 1/2 - 1/4**
S-2058	Sexauer or Skinner Tappg. Tools For use with Chicago, Economy,		**27 - 1/2 - 3/8**
S-2059	Sexauer or Skinner Tappg. Tools For use with Chicago, Economy,		**27 - 1/2 - 19/64**
S-2060	Sexauer or Skinner Tappg. Tools For use with Chicago, Economy,		**27 - 9/16 - 3/8**
S-2061	Sexauer or Skinner Tappg. Tools For use with Chicago, Economy,		**27 - 9/16 - 19/64**
S-2062	Sexauer or Skinner Tappg. Tools For use with Chicago, Economy,		**27 - 5/8 - 3/8**
S-2096	Assortment—1 each of 34 Refills		
S-2098	Asst. 81 different seats incl. seat and thread gauge		

CAP THREAD AND TOP BIBB GASKETS

A—Asbestos; R—Rubber; C.I.—Cloth Inserted Rubber; P—Plastic; F—Fibre; L—Leather

C-3200 C-3201 C-3202 C-3203 C-3204 C-3205 C-3205A C-3205B C-3205C C-3206 C-3206A C-3207 C-3207A C-3208 C-3209

Part No.	Use	O.D.	I.D.	Ht.	Matl.
C-3200	For Sterling and Gerber Diverter Valves				
C-3201	Am. Std. OEM 1319-07, Eljer, Gerber, Sterling, Wolverine	$1\frac{1}{32}$	$\frac{7}{8}$	$\frac{1}{8}$	A
C-3202	Rinse-Quick Spray Head and Hose Packing	$\frac{19}{32}$	$\frac{9}{32}$	$\frac{7}{32}$	R
C-3203	Am. Std. OEM 73-07	$1\frac{1}{16}$	$\frac{15}{16}$	$\frac{3}{32}$	R
C-3204	Fits Kohler, Savoy, Sayco, others	$\frac{7}{8}$	$\frac{3}{4}$	$\frac{1}{16}$	A
C-3205	Am. Std. OEM 142-07, Sears, Sterling, Yngstn.	$\frac{27}{32}$	$\frac{7}{16}$	$\frac{1}{16}$	A
C-3205A	Am. Std. OEM 125-07, Am. Kit., Harcraft, Repcal	$\frac{13}{16}$	$\frac{29}{64}$	$\frac{1}{16}$	A
C-3205B	Am. Brass, Sterling	$\frac{25}{32}$	$\frac{13}{32}$	$\frac{1}{16}$	A
C-3205C	Fits Briggs, Crane, Gerber, Speakman	$\frac{29}{32}$	$\frac{7}{16}$	$\frac{3}{32}$	C.I.
C-3206	Fits Repcal, Savoy, Union Brass	1	$\frac{51}{64}$	$\frac{1}{16}$	A
C-3206A	Fits Chicago, Gyro	$1\frac{3}{64}$	$\frac{51}{64}$	$\frac{3}{32}$	C.I.
C-3207	Fits various makes of faucets	$1\frac{3}{32}$	$\frac{53}{64}$	$\frac{1}{8}$	A
C-3207A	Crane OEM F10590	$1\frac{3}{32}$	$\frac{59}{64}$	$\frac{3}{32}$	Cork
C-3208	Fits Crane	$1\frac{3}{8}$	1	$\frac{1}{32}$	R
C-3209	Diamond Shape to fit Crane				

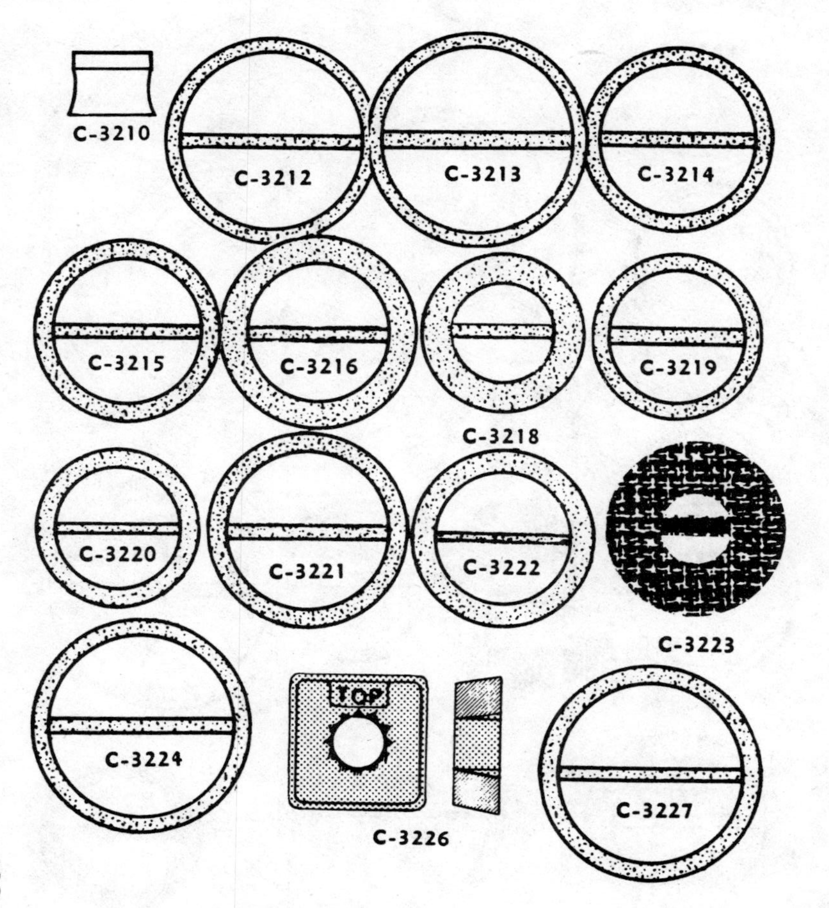

C-3210	OEM Delta Seat (used with L-2611)				
C-3212	Eljer, Gerber, Kohler, Milwaukee, Speakman	1 1/16	15/16	1/16	A
C-3213	Fits various makes	1 7/32	1 1/32	1/32	A
C-3214	Fits Gerber OEM 33, Am. Std., Speakman, Union Brass	31/32	13/16	1/16	A
C-3215	Fits various makes	15/16	3/4	1/16	A
C-3216	Fits Kohler, Speakman	1 3/64	3/4	1/16	A
C-3218	Fits Kohler, Moen	13/16	1/2	1/16	A
C-3219	Fits Sayco	7/8	11/16	3/32	A
C-3220	Fits Am. Kitchen, Indiana Brass	13/16	5/8	1/32	A
C-3221	Fits Chicago, Speakman, Sayco	1 1/32	7/8	1/16	A
C-3222	Fits Union Brass	15/16	43/64	1/32	A
C-3223	Fits Briggs, Kohler Repcal, Sears, Sterling, Wolverine	15/16	3/8	3/32	C.I.
C-3224	Fits Briggs, Kohler	1 1/8	31/32	1/16	A
C-3226	Handle Insulator to fit Crane OEM F12560				
C-3227	Fits Am. Std., Wolverine	1 1/8	15/16	1/16	A

CAP THREAD AND TOP BIBB GASKETS

A—Asbestos; R—Rubber; C.I.—Cloth Inserted Rubber; P—Plastic; F—Fibre; L—Leather

Part No.	Use	O.D.	I.D.	Ht.	Matl.
C-3228	Fits Am. Std. CEM 12007-07, Schaible, Sears, Youngstown Single Lever Faucets	13/16	5/8	1/16	R
C-3229	Fits Crane, Union Brass	1 5/32	1	1/32	A
C-3230	Fits various makes	15/16	21/32	1/16	A
C-3231	Fits Repcal, Sterling	1 1/4	1 1/32	1/16	A
C-3232	Fits Crane, Gerber, Price-Pfister, Savoy, Sayco, Sterling, Union Brass	27/32	3/4	1/16	A
C-3233	Fits various makes	1 1/32	23/64	3/32	C.I.
C-3237	Fits Gerber, Indiana Brass, Kohler, Sterling	29/32	13/16	1/16	A
C-3238	Delta Cap Gasket used with P-2232 (OEM 61A)				
C-3238A	Delta Cap Gasket used with P-2232A				
C-3239	Fits Am. Std. OEM 1044-07	1	27/32	9/32	R
C-3242	Fits Am. Std. OEM 998-07	2 45/64	2 29/64	1/16	A
C-3243	Fits Am. Std. OEM 1325-07	37/64	25/64	1/16	A
C-3244	Fits Am. Std. OEM 1326-07	9/16	17/64	5/64	A
C-3247	Fits Harcraft	55/64	49/64	1/16	F

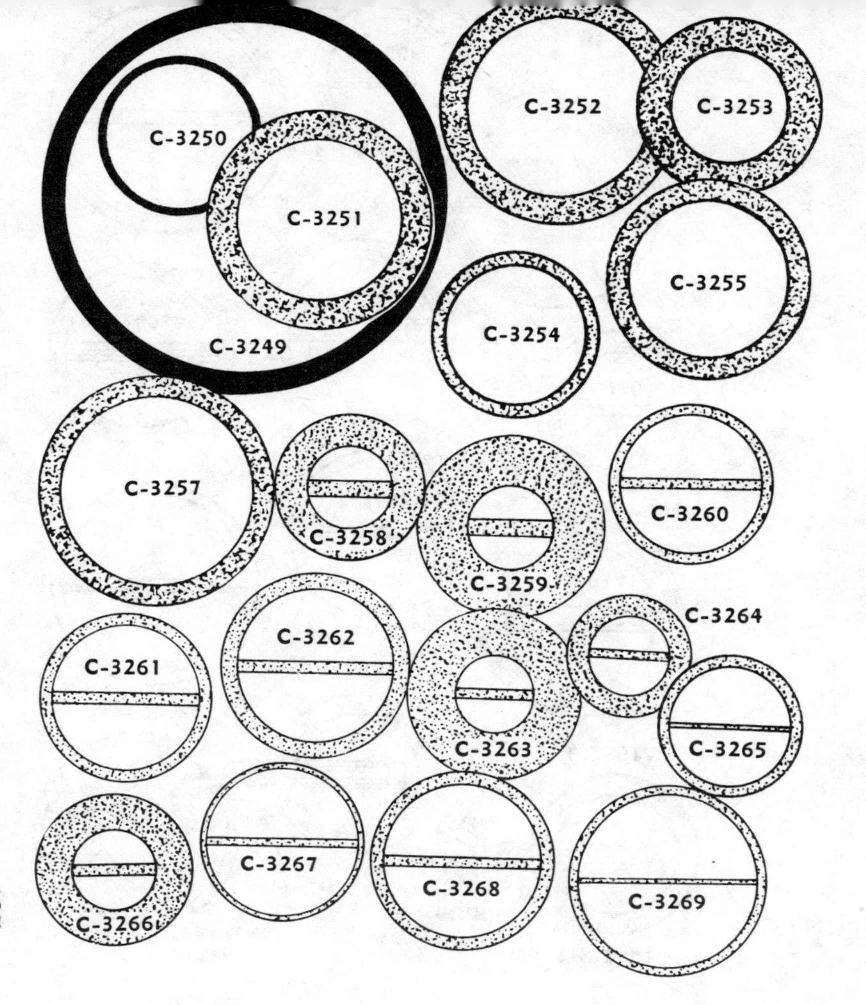

C-3249	Fits Kohler Niedecken Mixer	2¼	1 31/32	1/16	R
C-3250	Fits Price-Pfister	27/32	¾	1/16	P
C-3251	Fits Price-Pfister	1 3/16	⅞	1/16	P
C-3252	Fits Price-Pfister	1 15/64	1 1/16	1/16	P
C-3253	Fits Price-Pfister	15/16	39/64	1/16	P
C-3254	Fits Repcal	⅞	¾	1/16	F
C-3255	Fits Speakman	1 1/16	57/64	1/32	F
C-3257	Fits Am. Std. OEM 408-17	1 1/32	1	⅛	R
C-3258	Fits Am. Brass	25/32	7/16	3/32	F
C-3259	Fits Briggs, Empire	63/64	7/16	3/32	F
C-3260	Fits Am. Brass, Sayco, Union Brass	55/64	¾	1/16	F
C-3261	Fits Savoy, Sayco	57/64	49/64	1/16	F
C-3262	Fits Chicago	1	13/16	1/16	F
C-3263	Fits Barnes, Eljer	29/32	13/32	1/16	F
C-3264	Fits Eljer	21/32	13/32	3/64	F
C-3265	Fits Price-Pfister, Schaible	¾	⅝	1/32	F
C-3266	Fits Am. Std.	13/16	7/16	1/16	F
C-3267	Fits Harcraft, Price-Pfister	27/32	25/32	1/16	F
C-3268	Fits Gerber, Repcal Union Brass	31/32	13/16	1/16	F
C-3269	Fits Eljer	1 1/32	15/16	1/32	F

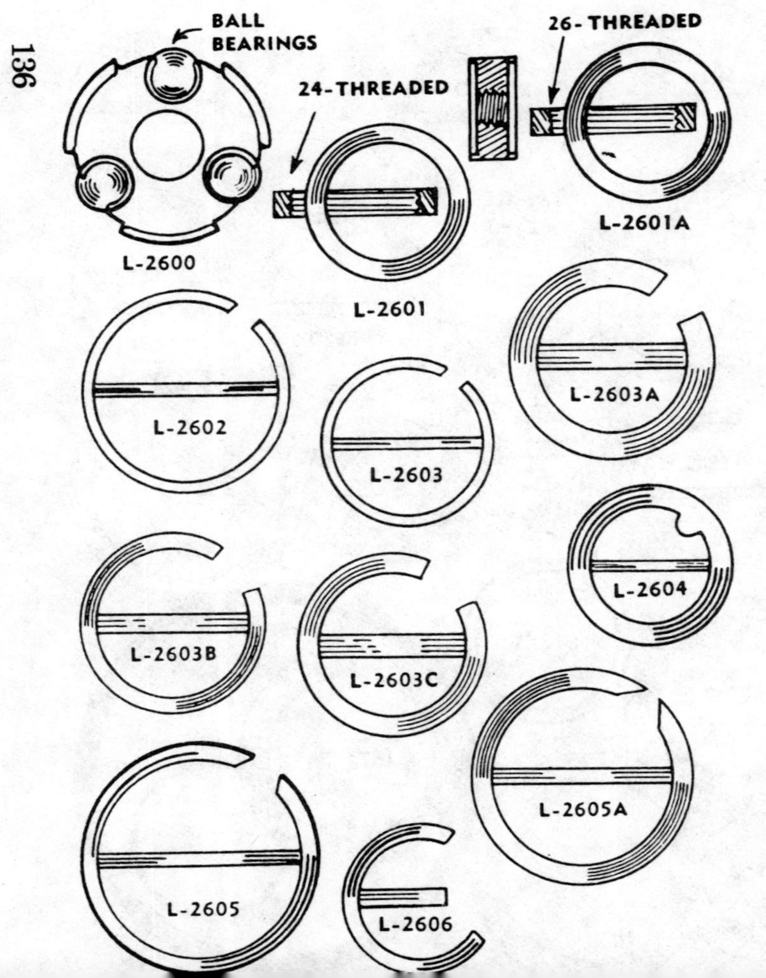

LOCK RINGS & MISC. FAUCET PARTS

Part No.	Use	O.D.	I.D.	Ht.	Shape
L-2600	Am. Std. and Kohler Ball Race for Self-Closing Faucets. Am. Std. OEM 527-17, Kohler OEM 31537				
L-2601	Fits Am. Brass,	55/64	21/32	9/64	24Th
L-2601A	Fits Speakman	55/64	21/32	9/64	26Th
L-2602	Fits Am. Std., Am. Kit., Crosley, Eljer, Repcal, Sears, Youngstown	1 1/8	1	1/16	S
L-2603	Fits Am. Kit., Barnes Price-Pf., Repcal, Schaible, Sears, Tracy, Youngstown	29/32	25/32	1/16	S
L-2603A	Fits various makes	1 1/16	25/32	1/8	S
L-2603B	Fits Barnes, Briggs, Crane, Universal-Rundle, Youngstown	31/32	13/16	5/64	Rnd
L-2603C	Fits Crane	63/64	49/64	7/64	Sq
L-2604	Am. Std. Lock Washer OEM 21365-05				
L-2605	Fits Am. Std., Repcal	1 1/4	1 1/16	5/64	S
L-2605A	Fits Am. Std. OEM 684-27	1 5/32	1	5/64	
L-2606	Fits Am. Brass, Indiana Brass, Repcal	53/64	41/64	3/32	S

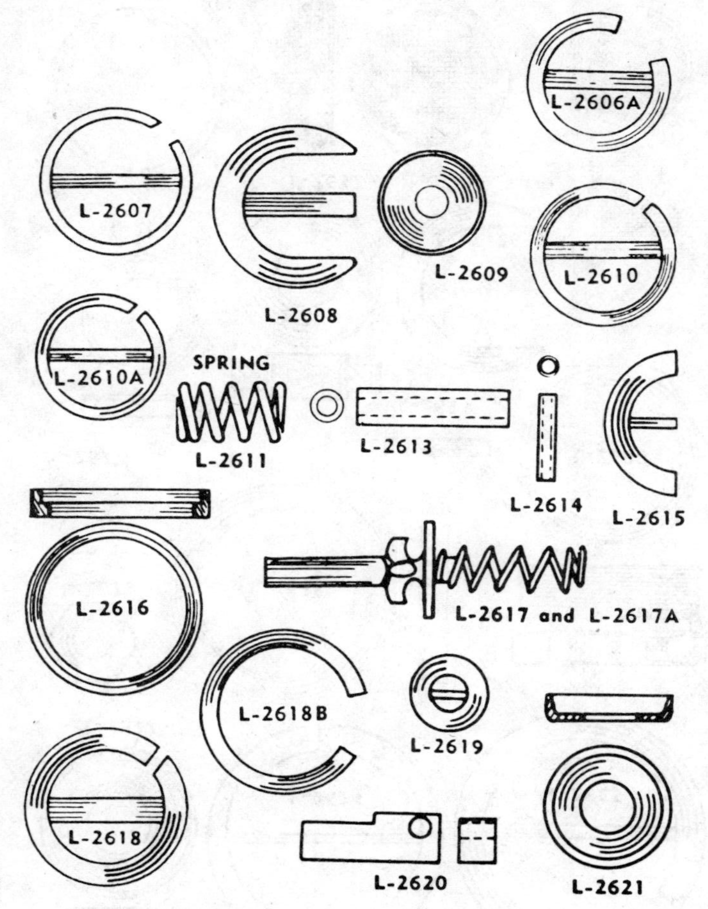

L-2606A	Fits Am. Brass, Central Br., Union Br.	47/64	37/64	5/64	Rnd
L-2607	Fits Barnes, Central, Crane, Dick, Eljer, Milw., Schaible, Sears, Sterling, Youngstown	37/64	45/64	1/16	S
L-2608	Central OEM 72F Spout Ring Horseshoe				
L-2609	Central Transfer				Th
L-2610	Fitis Crane, Price-Pf., Union Brass	3/4	5/8	1/16	S
L-2610A	Fits Crane, Dick Bros., Gerber	11/16	9/16	1/16	S
L-2611	Delta Monel Spring used with C-3210 (part of OEM 134A)				
L-2613	Moen OEM 473 Handle Pin	13/16			
L-2614	Moen OEM 474 Anchor Pin	1/2			
L-2615	Am. Std. OEM 401-27	25/32			
L-2616	Fits Briggs, Schaible, Youngstown	29/32	25/32	1/8	
L-2617	Am. Std. Single Lever Plunger and Spring OEM 12270-07, 12179-07				
L-2617A	Spring for L-2617 Am. Std. OEM 12179-07				
L-2618	Sterling OEM A-10X	7/8	21/32	7/64	S
L-2618B	Chicago Lockring OEM 1-26				
L-2619	Fits Chicago OEM 1-31	7/16	7/32	1/32	
L-2620	Moen Anchor OEM 005	23/32			
L-2621	Chicago Monel Retainer OEM 1-22				

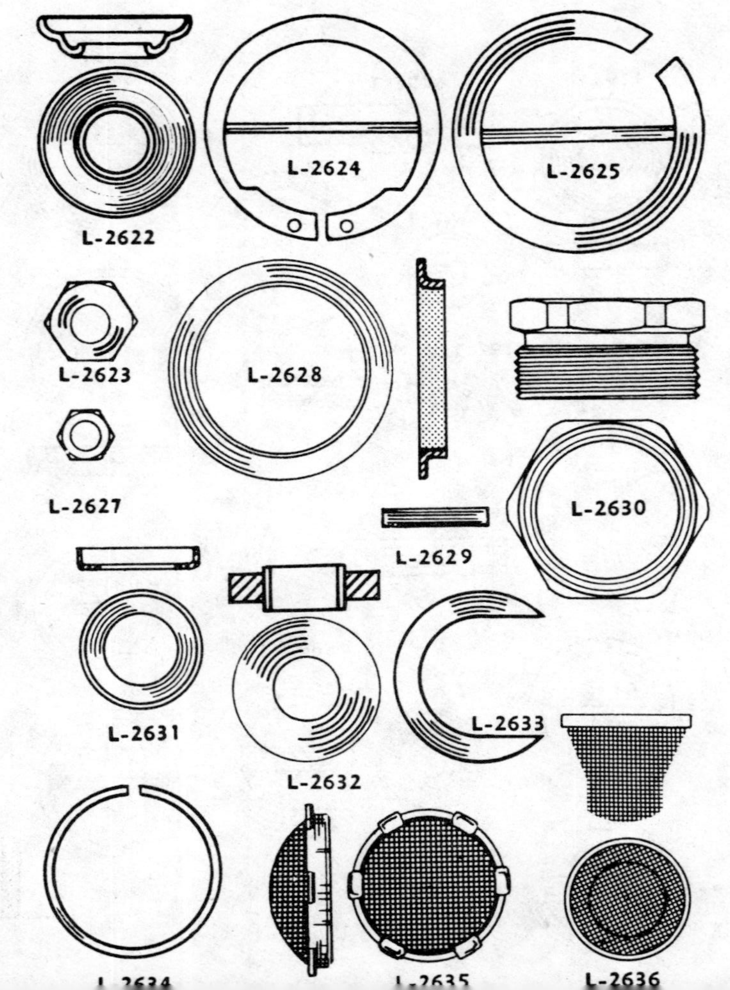

LOCK RINGS & MISC. FAUCET PARTS

Part No.	Use	O.D.	I.D.	Ht.	Shape
L-2622	Chicago Monel Seat OEM 1-27				
L-2623	Chicago Stem Nut OEM 1-18				
L-2624	Gyro OEM 9-C5 Snap Ring	1 9/32			
L-2625	Fits Eljer	1 19/64	1 1/32	1/16	S
L-2627	Hex Nut	1/4	10-32		
L-2628	Gyro Handle Bearing OEM 9-30A				
L-2629	Gyro Handle Pin OEM 9-C3				
L-2630	Locknut for Am. Std. Aqua Seal OEM 855-17				
L-2631	Milwaukee OEM P3342 Monel Ret. Cup	19/32	13/32	7/64	
L-2632	Am. Std. Colony Packing OEM 72102-07				
L-2633	Harcraft Lock Washer OEM 1568				
L-2634	Harcraft OEM 1993	15/16	13/16	1/16	
L-2635	Am. Std. Cone Shaped Screen OEM 72507-07				
L-2636	Am. Std. Cone Shaped Screen OEM 12269-07				
L-2692	100 Asst.				
L-2699	41 X-Refills Asst.				

SWING SPOUT PACKINGS (ALL REG. WALLS)

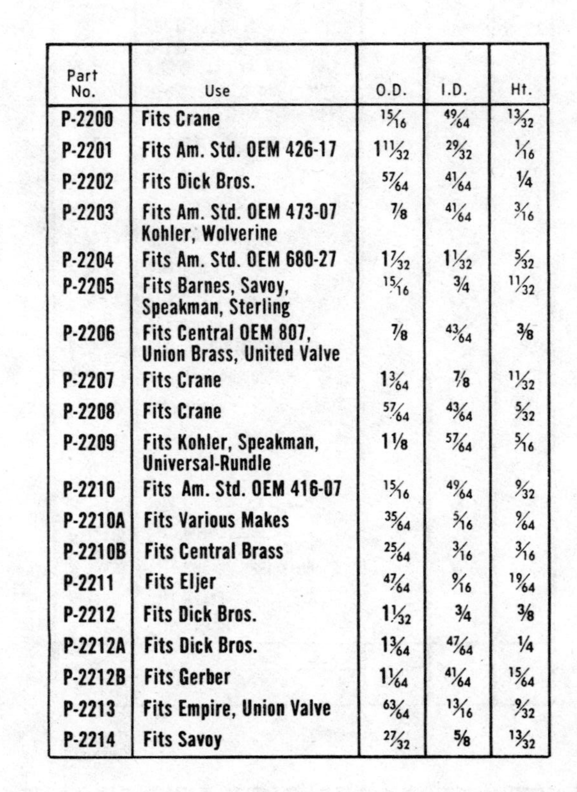

Part No.	Use	O.D.	I.D.	Ht.
P-2200	Fits Crane	15/16	49/64	13/32
P-2201	Fits Am. Std. OEM 426-17	1 11/32	29/32	1/16
P-2202	Fits Dick Bros.	57/64	41/64	1/4
P-2203	Fits Am. Std. OEM 473-07 Kohler, Wolverine	7/8	41/64	3/16
P-2204	Fits Am. Std. OEM 680-27	1 7/32	1 1/32	5/32
P-2205	Fits Barnes, Savoy, Speakman, Sterling	15/16	3/4	11/32
P-2206	Fits Central OEM 807, Union Brass, United Valve	7/8	43/64	3/8
P-2207	Fits Crane	1 3/64	7/8	11/32
P-2208	Fits Crane	57/64	43/64	5/32
P-2209	Fits Kohler, Speakman, Universal-Rundle	1 1/8	57/64	5/16
P-2210	Fits Am. Std. OEM 416-07	15/16	49/64	9/32
P-2210A	Fits Various Makes	35/64	5/16	9/64
P-2210B	Fits Central Brass	25/64	3/16	3/16
P-2211	Fits Eljer	47/64	9/16	19/64
P-2212	Fits Dick Bros.	1 1/32	3/4	3/8
P-2212A	Fits Dick Bros.	1 3/64	47/64	1/4
P-2212B	Fits Gerber	1 1/64	41/64	15/64
P-2213	Fits Empire, Union Valve	63/64	13/16	9/32
P-2214	Fits Savoy	27/32	5/8	13/32

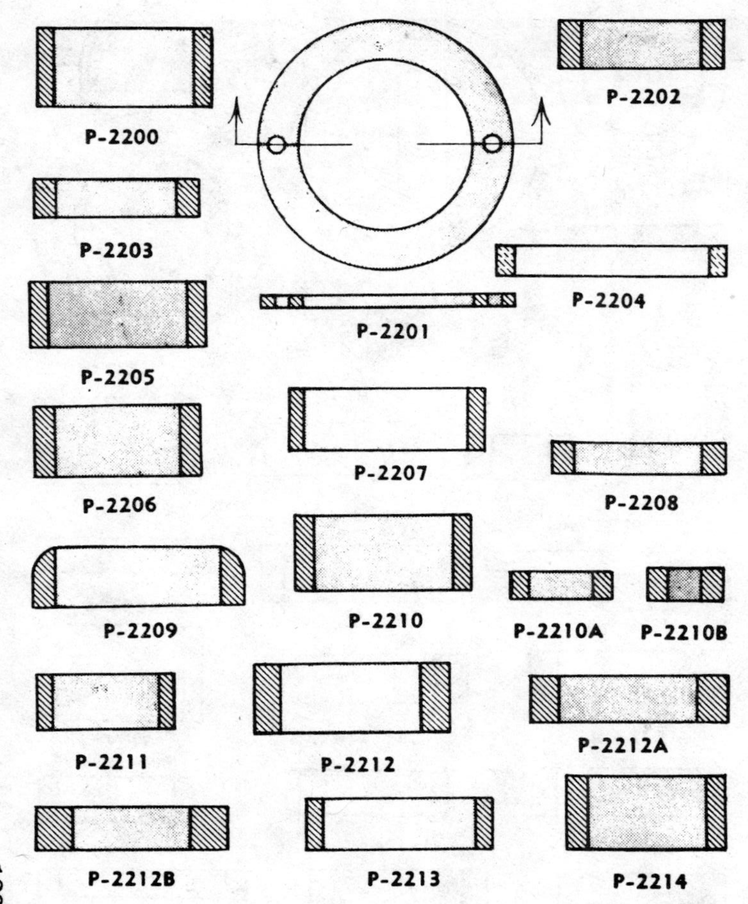

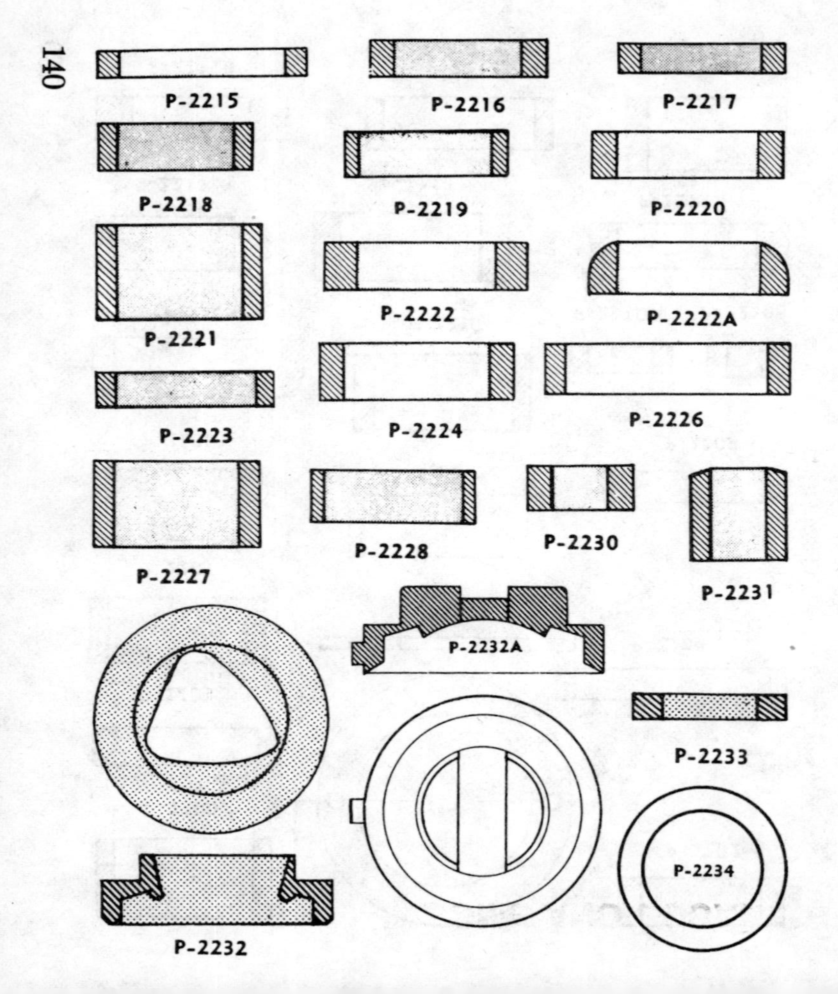

SWING SPOUT PACKINGS (ALL REG. WALLS)

Part No.	Use	O.D.	I.D.	Ht.
P-2215	Fits Kohler	17/64	57/64	5/32
P-2216	Fits Kohler	15/16	11/16	13/64
P-2217	Fits Kohler, Wolverine	7/8	41/64	5/32
P-2218	Fits Milwaukee	13/16	5/8	17/64
P-2219	Fits Briggs, Sterling	55/64	45/64	1/4
P-2220	Fits Price-Pfister	1	3/4	17/64
P-2221	Fits Am. Brass	7/8	43/64	17/32
P-2222	Fits Speakman	1 5/64	3/4	17/64
P-2222A	Fits Speakman	1 1/16	49/64	9/32
P-2223	Fits Harcraft, Schaible, Speakman	15/16	3/4	3/16
P-2224	Fits Sears	1 1/32	25/32	5/16
P-2226	Fits Eljer	1 19/64	1 5/64	9/32
P-2227	Fits Empire, Repcal	7/8	43/64	15/32
P-2228	Fits Chicago, Repcal	7/8	3/4	9/32
P-2230	Fits Kohler	9/16	19/64	15/64
P-2231	Fits Gyro OEM 9C4	33/64	5/16	1/2 Bv
P-2232	Fits Delta Cam Assy. OEM 61A used with our C-3238			
P-2232A	Fits Delta Cam Assy. for Del-Dial only. Used with our C-3238A			
P-2233	Fits Harcraft	25/32	31/64	9/64
P-2234	Sterling OEM 99S-0090 Spout Packing			

DOME AND SQUARE STEM BONNET PACKING

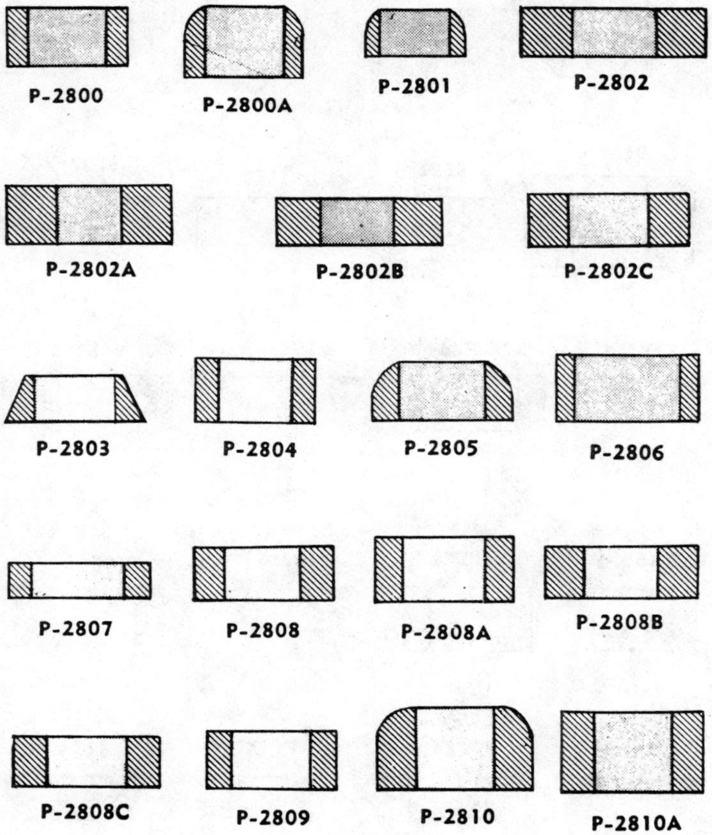

P-2800 P-2800A P-2801 P-2802

P-2802A P-2802B P-2802C

P-2803 P-2804 P-2805 P-2806

P-2807 P-2808 P-2808A P-2808B

P-2808C P-2809 P-2810 P-2810A

"S"—SQUARE "D"—DOME

Part No.	Use	O.D.	I.D.	Ht.	Shape
P-2800	Fits Am. Std. OEM 397 Renu Valves, Barnes, Union Brass	41/64	27/64	5/16	S
P-2800A	Fits Am. Brass, Empire, United Valve	41/64	27/64	3/8	D
P-2801	Fits Am. Brass	17/32	13/32	1/4	D
P-2802	Fits Youngstown	63/64	29/64	1/4	S
P-2802A	Fits various makes	7/8	11/32	5/16	S
P-2802B	Fits Am. Kit., Schaible	7/8	25/64	1/4	S
P-2802C	Fits Price-Pfister, Sterling, Tracy	55/64	7/16	9/32	S
P-2803	Fits Barnes, Sterling	49/64	27/64	1/4	Bev
P-2804	Fits Briggs, Central, Gerber, Harcraft, Repcal, Sterling, Wolv.	41/64	25/64	11/32	S
P-2805	Fits Speakman	3/4	29/64	5/16	D
P-2806	Fits Am. Std. OEM 679-27	3/4	19/32	11/32	S
P-2707	Fits Barnes	3/4	1/2	3/16	S
P-2808	Fits Briggs, Schaible	3/4	13/32	9/32	S
P-2808A	Fits Harcraft, Kohler, Repcal	47/64	7/16	11/32	S
P-2808B	Fits Barnes, Central, Crane, Univ.-Rundle	13/16	13/32	9/32	S
P-2808C	Fits Central	25/32	27/64	9/32	S
P-2809	Fits Eljer, Harcraft, Repcal	11/16	27/64	5/16	S
P-2810	Fits Am. Std., Gerber, Repcal	53/64	13/32	29/64	D
P-2810A	Fits Briggs, Eljer	3/4	27/64	7/16	S

DOME AND SQUARE STEM BONNET PACKING

"S"—SQUARE "D"—DOME

P-2811 P-2811A P-2811B P-2812

P-2813 P-2815 P-2816 P-2817

P-2818 P-2819 P-2820 P-2822

P-2823 P-2824 P-2825 P-2826

P-2827 P-2828 P-2829 P-2830

Part No.	Use	O.D.	I.D.	Ht.	Shape
P-2811	Fits Briggs	45/64	13/32	1/4	D
P-2811A	Fits Speakman, United Valve	39/64	7/16	3/16	S
P-2811B	Fits Central, Union Brass	23/32	27/64	15/64	Bev
P-2812	Fits various makes	41/64	27/64	7/64	S
P-2813	Fits Am. Std., Yngstwn. OEM 54111, Crosley, Tracy	1	7/16	5/16	S
P-2815	Fits Chicago	43/64	13/32	15/64	D
P-2816	Fits Empire, Gerber, Price-Pf., Union Brass	43/64	13/32	11/32	S
P-2817	Fits Eljer	11/16	27/64	11/32	D
P-2818	Fits Kohler, Repcal	57/64	29/64	13/32	S
P-2819	Fits Kohler	47/64	7/16	11/64	D
P-2820	Fits Crane, Milwkee., Speakman, Sterling	47/64	7/16	5/16	D
P-2822	Fits Savoy, Sayco, Union Brass	9/16	3/8	15/64	S
P-2823	Fits Am. Std. and Schaible OEM 57283, Epeakman OEM 49-92	3/4	25/64	1/4	S
P-2824	Fits Am. Std. OEM 397, Harcraft, Speakman, Sterling	5/8	3/8	1/4	S
P-2825	Fits Price-Pfister. Repcal, Wolverine	23/32	13/32	11/32	D
P-2826	Fits various makes	29/32	13/32	31/64	S
P-2827	Fits various makes	7/8	7/16	3/8	D
P-2828	Fits various makes	21/32	27/64	21/64	Bev
P-2829	Fits Am. Std. OEM 1311-07, Dick, Sears	47/64	25/64	3/8	D
P-2830	Fits Sterling OEM 99S-0509				

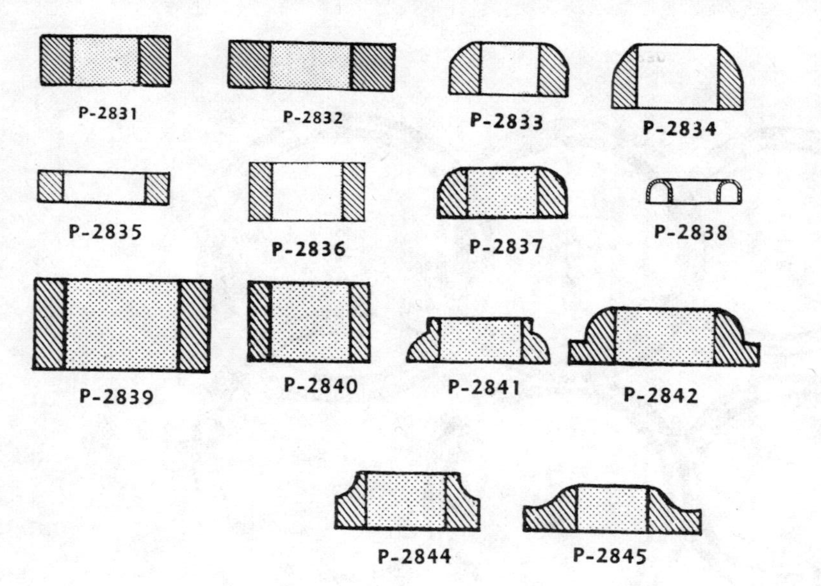

P-2831	Fits Sterling OEM 99S-0386				
P-2832	Fits Sterling OEM 99S-0028				
P-2833	Fits various makes	5/8	5/16	9/32	D
P-2834	Fits Am. Std.	11/16	7/16	11/32	D
P-2835	Fits Kohler	11/16	7/16	5/32	S
P-2836	Fits Briggs, Savoy, Sayco, Union Brass	39/64	3/8	5/16	S
P-2838	"V" Packing for Am. Std. Push-Pull Lav. Single Lev. Fct. OEM 29312-07				
P-2839	Fits Am. Std. OEM 391-07	29/32	9/16	15/32	S
P-2840	Fits Briggs	19/32	25/64	25/64	S
P-2841	Fits Kohler	3/4	7/16	7/32	D
P-2842	Fits Speakman	1	7/16	11/32	D
P-2844	Fits Sterling OEM 99S-0427	49/64	25/64	5/16	D
P-2845	Fits Union Brass OEM 1903	15/16	23/64	9/32	D

"O" RINGS

PACKING RINGS FOR SWING SPOUT, SLIP-JOINTS, PUMP AND HYDRAULIC APPLICATIONS

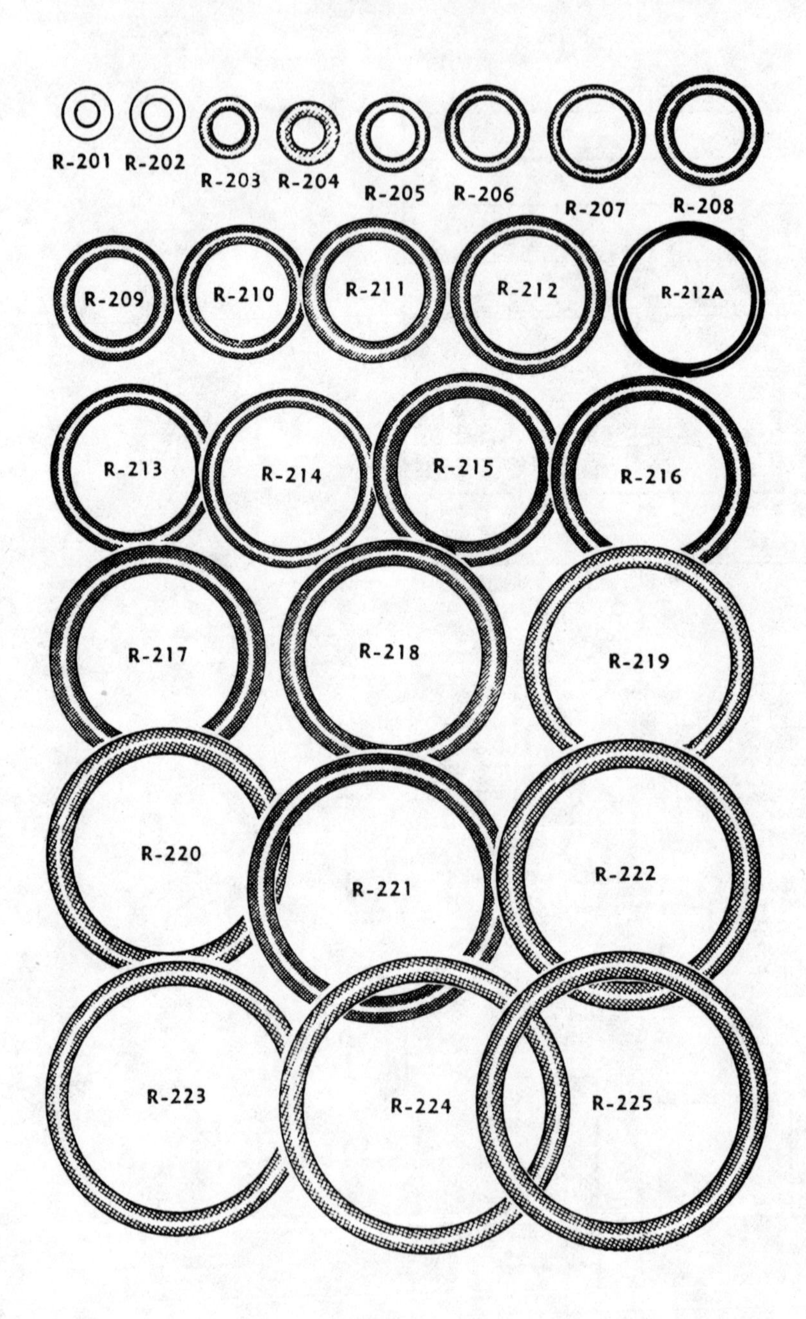

Part No.	Use	O.D.	I.D.	Wall
R-201		1/4	1/8	1/16
R-202		9/32	5/32	1/16
R-203		5/16	3/16	1/16
R-204		11/32	7/32	1/16
R-205	Fits Central, Kohler, Schaible, Tracy	3/8	1/4	1/16
R-206	Fits Central, Crane, ElJer, Kohler, Milwaukee, Sears, Sterling	7/16	5/16	1/16
R-207	Fits Crane, Eljer, Milwaukee. Speakman, Universal-Rundle, Elkay OEM 30453	1/2	3/8	1/16
R-208	Fits Am. Kit., Crane, Gerber, Harcraft, Indiana Brass, Milwaukee, Price-Pfister, Repcal, Sterling, Sayco. Union Brass	9/16	3/8	3/32
R-209	Fits Gyro, Kohler OEM 34263, Moen, Sears, Savoy	5/8	7/16	3/32
R-210	Fits Central, Gerber, Indiana Brass. Moen, Univ. Rundle	11/16	1/2	3/32
R-211	Fits Am. Std., Barnes, Central, Crane. Dick, Eljer, Indiana Brass, Milwaukee, Repcal, Moen, Price-Pfister. Schaible, Sears, Sterling	3/4	9/16	3/32
R-212	Fits Am. Std., Moen, Speakman, Union Brass	13/16	5/8	3/32
R-212A	Fits various makes	13/16	11/16	1/16
R-213	Fits Am. Kit., Barnes, Briggs, Gyro, Indiana Brass, Kohler, Moen, Price-Pfister, Repcal, Schaible, Sears, Tracy, Union Brass, Youngstown	7/8	11/16	3/32
R-214	Fits Crane	15/16	3/4	3/32
R-215	Fits Crane	1	3/4	1/8
R-216	Fits Eljer, Gyro, Moen	1 1/16	13/16	1/8
R-217	Fits Am. Std. OEM 507-37, Moen 137, Repcal	1 1/8	7/8	1/8
R-218	Fits Speakman	1 3/16	15/16	1/8
R-219	Fits various makes	1 1/4	1	1/8
R-220	Fits various makes	1 5/16	1 1/16	1/8
R-221	Fits Moen Model 52	1 3/8	1 1/8	1/8
R-222	Fits various makes	1 7/16	1 3/16	1/8
R-223	Fits Sears, Moen Model 42E	1 1/2	1 1/4	1/8
R-224	For special applications	1 9/16	1 5/16	1/4
R-225	For special applications	1 5/8	1 3/8	1/8

"O" RINGS

PACKING RINGS FOR SWING SPOUT, SLIP-JOINTS, PUMP AND HYDRAULIC APPLICATIONS

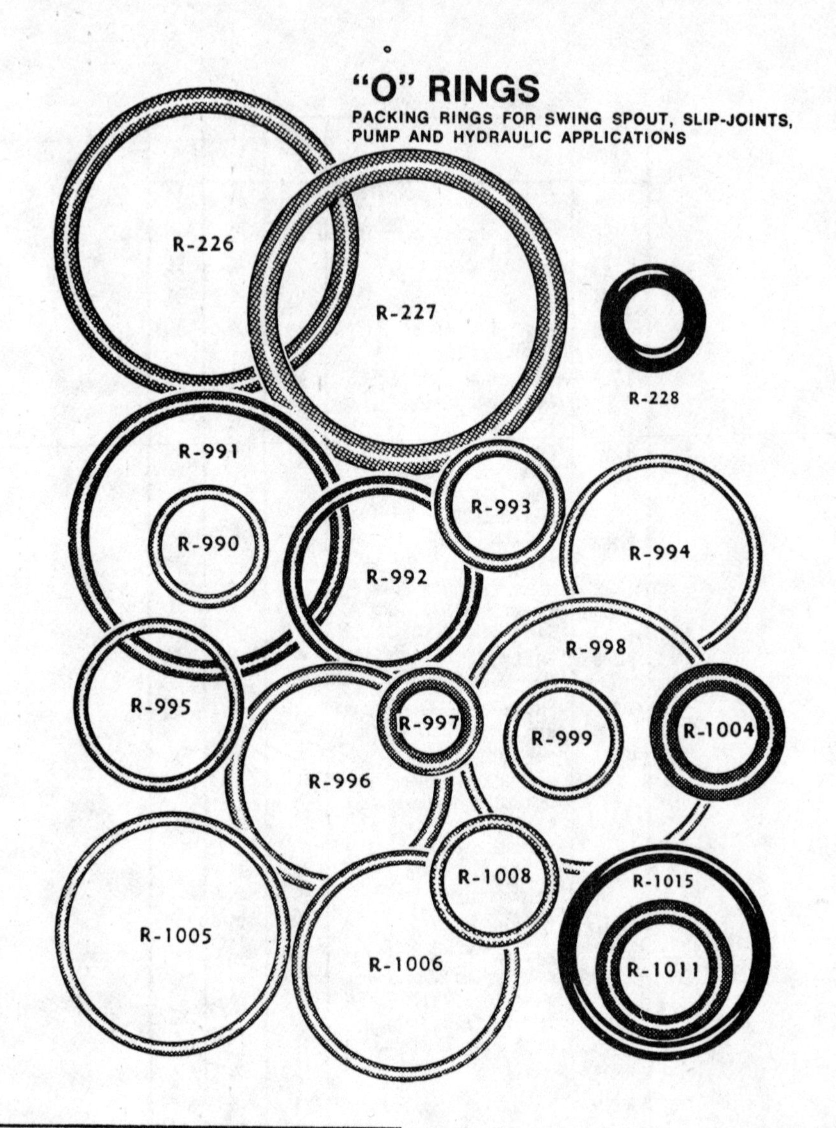

Part No.	Use	O.D.	I.D.	Wall
R-992	Fits Am. Std. OEM 55234, Am. Kit., Crane, Crosley, Eljer, Kohler, Repcal, Schaible, Sears, Tracy, Yngstwn., Kohler OEM 29464	1 1/16	7/8	3/32
R-993	Fits Central, Gerber, Moen, Indiana Brass, Univ.-Rundle	11/16	1/2	3/32
R-994	Fits Kohler OEM 34264	1 1/16	15/16	1/16
R-995	Fits Am. Std., Kohler OEM 34300	7/8	3/4	1/16
R-996	Fits Alamark	1 3/16	1	3/32
R-997	Fits Eljer OEM 5290	9/16	5/16	1/8
R-998	Fits Moen Model 42E	1 7/16	1 5/16	1/16
R-999	Fits Am. Std. OEM 886-17	5/8	1/2	1/16
R-1004	Fits Crane OEM F12376	45/64	27/64	9/64
R-1005	Fits Delta	1 1/4	1 1/8	1/16
R-1006	Fits Am. Std. OEM 12035, Schaible, Sears, Youngstown	1 3/16	1 1/16	1/16
R-1008	Fits Harcraft	11/16	9/16	1/16
R-1011	Fits Barnes	23/32	15/32	9/64
R-1015	Fits Eljer OEM 5285 Lusterline, Burlington 68-9	1 1/4	1 1/16	3/32

BRASS FRICTION RINGS
SUPPLY AND WASTE

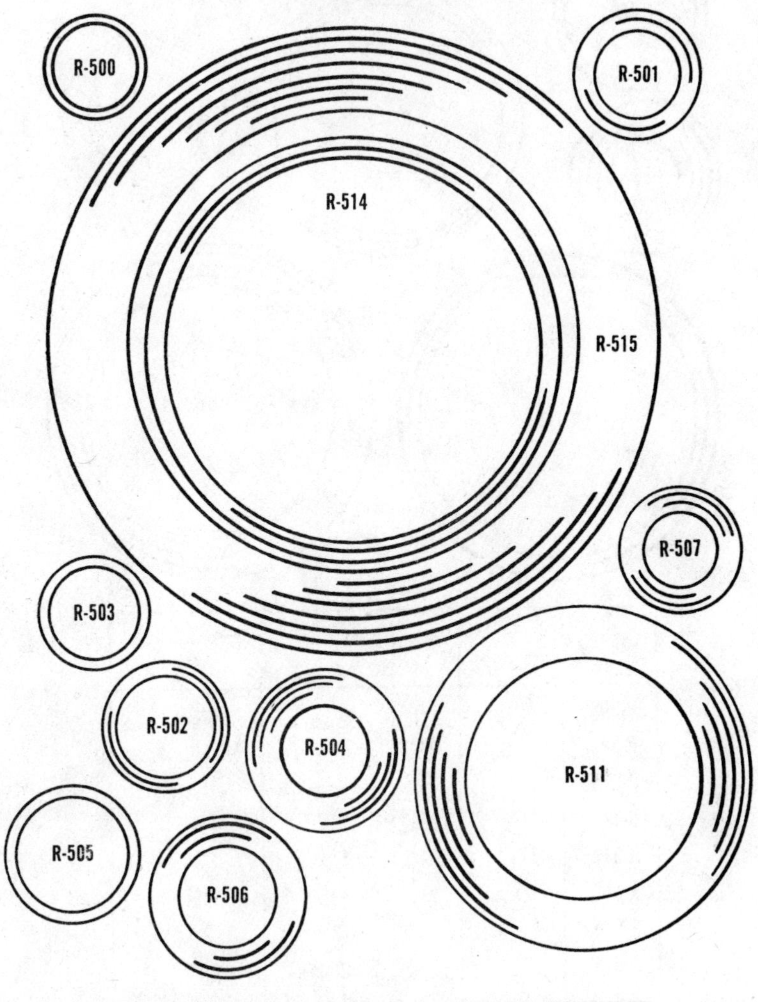

Part No.	I.D.	O.D.	Tube Size	Nut Size
R-500	13/32	1/2	3/8	FLEX. SUPPLY
R-501	29/64	21/32	7/16	1/2
R-502	33/64	21/32	1/2	1/2
R-503	29/64	9/16	7/16	3/8
R-504	15/32	13/16	7/16	BALLCOCK
R-505	19/32	45/64	9/16	1/2" I.P.
R-506	17/32	13/16	1/2	BALLCOCK
R-507	25/64	21/32	3/8	1/2

BRASS FRICTION RINGS

SUPPLY AND WASTE

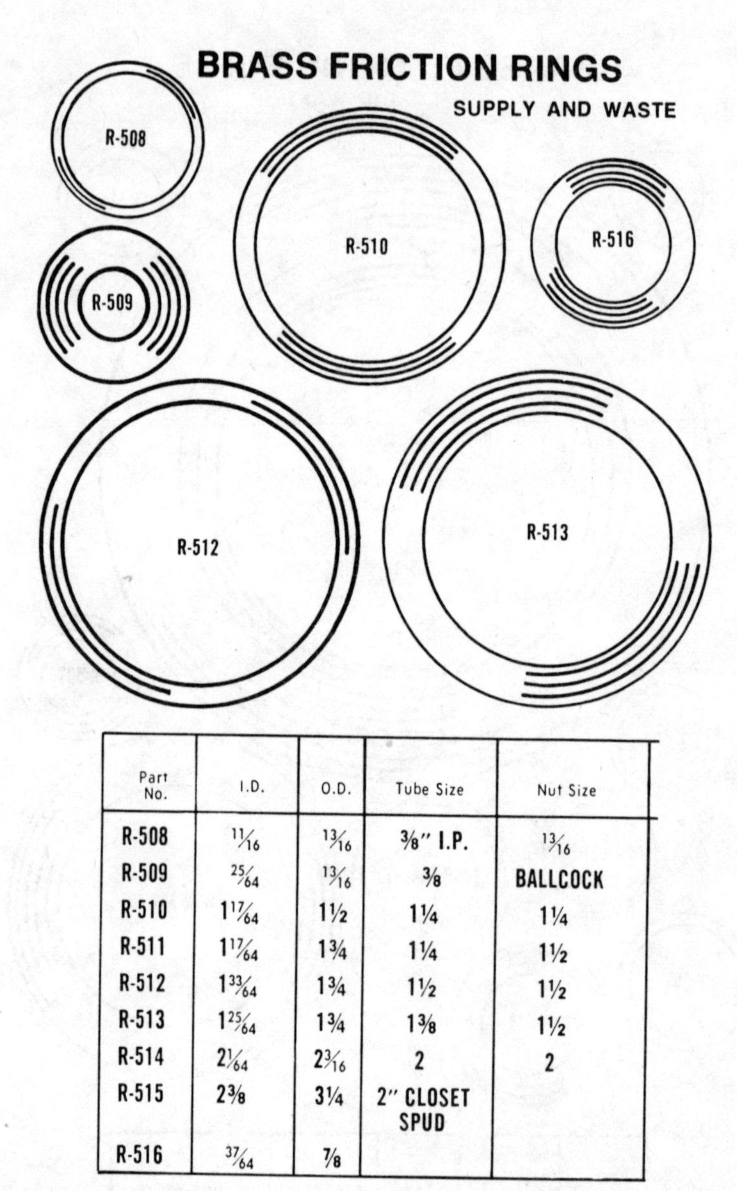

Part No.	I.D.	O.D.	Tube Size	Nut Size
R-508	11/16	13/16	3/8" I.P.	13/16
R-509	25/64	13/16	3/8	BALLCOCK
R-510	1 17/64	1 1/2	1 1/4	1 1/4
R-511	1 17/64	1 3/4	1 1/4	1 1/2
R-512	1 33/64	1 3/4	1 1/2	1 1/2
R-513	1 25/64	1 3/4	1 3/8	1 1/2
R-514	2 1/64	2 3/16	2	2
R-515	2 3/8	3 1/4	2" CLOSET SPUD	
R-516	37/64	7/8		

These brass friction rings may be used with our slip-joint washers. The following chart gives the size ring to be used with each washer:

SLIP JOINT WASHER	BRASS FRICTION RING	SLIP JOINT WASHER	BRASS FRICTION RING
W-500	R-506	W-506	R-502
W-501	R-501	W-507	R-508
W-502	R-501	W-508	R-508
W-503	R-502	W-510	R-509
W-504	R-503	W-511	R-500
W-505	R-506		

FRICTION RINGS

PACKING RINGS FOR SWING SPOUT, SLIP-JOINTS, PUMP AND HYDRAULIC APPLICATIONS

R-2400

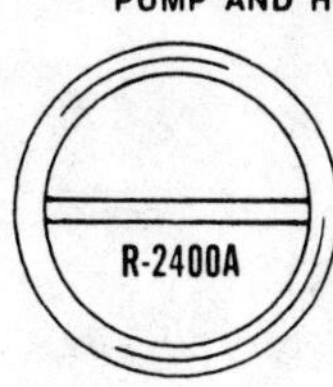

R-2400A

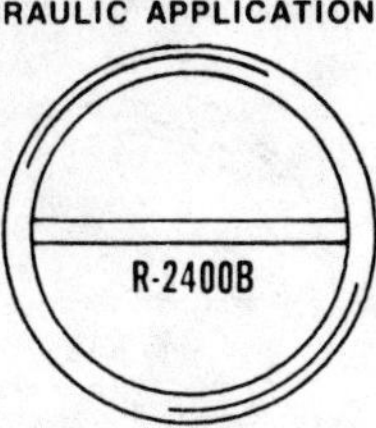

R-2400B

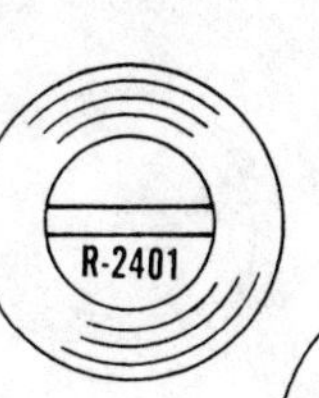

R-2401

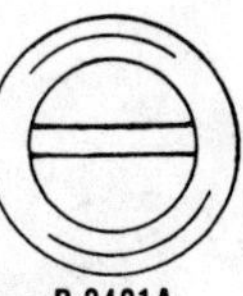

R-2401A

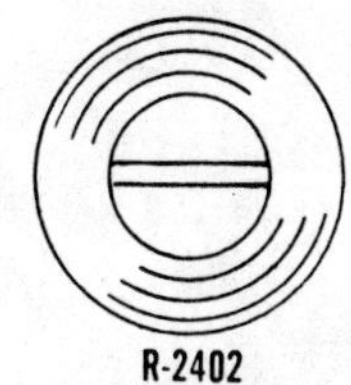

R-2402

R-2406A

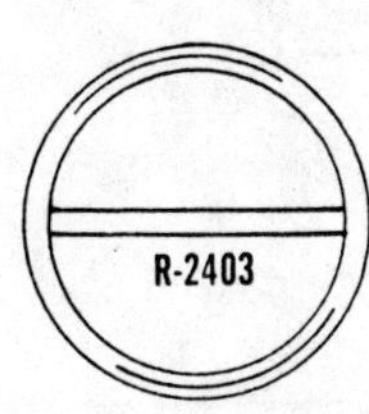

R-2403

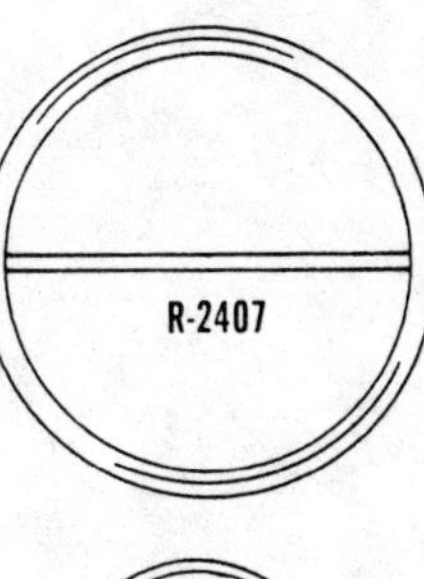

R-2407

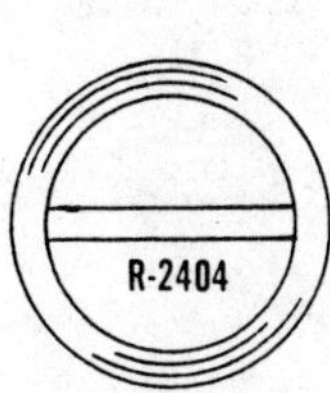

R-2404

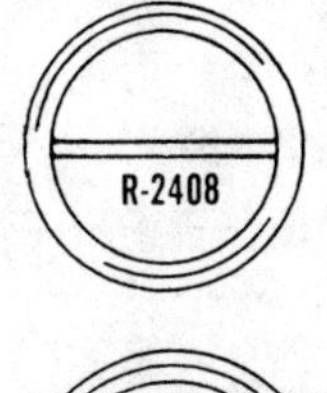

R-2408

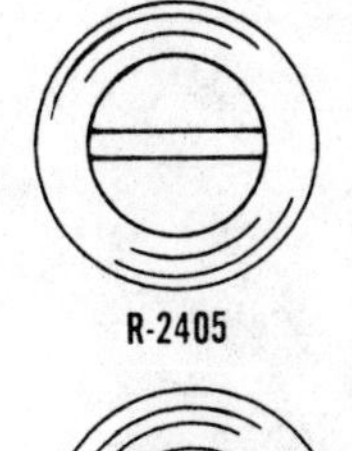

R-2405

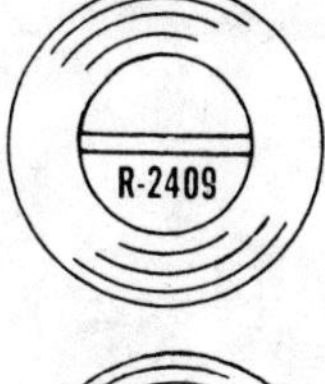

R-2409

R-2405A

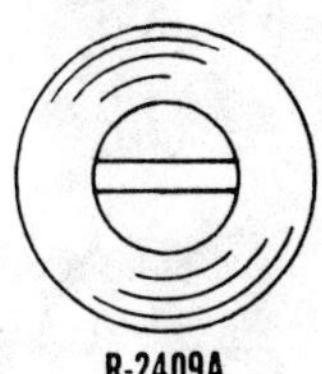

R-2409A

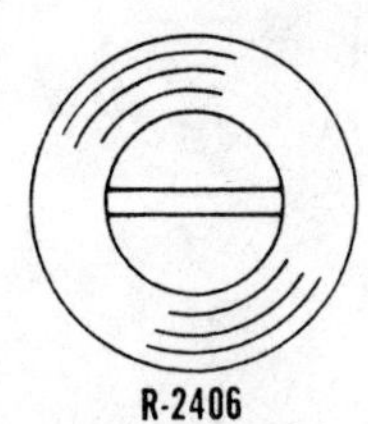

R-2406

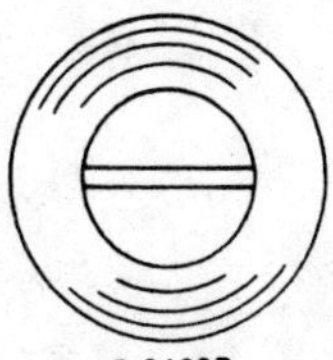

R-2409B

Part No.	Use	O.D.	I.D.	Ht.
R-2400	Fits Am. Brass, Central, Empire, Sterling, Savoy	55/64	11/16	5/64
R-2400A	Fits Central	7/8	43/64	1/16
R-2400B	Fits United Valve	63/64	53/64	1/16
R-2401	Fits Am. Std., Barnes Crane, Gerber, Kohler, Repcal, Sears, Sterling	53/64	29/64	5/64
R-2401A	Fits Am. Std. OEM 21625-07, Speakman, Sterling	11/16	7/16	5/64
R-2402	Fits Am. Brass. Chicago, Eljer, Empire, United Valve, Universal-Rundle	13/16	27/64	1/16
R-2403	Fits Am. Std., Barnes, Gyro, Kohler, Speakman	59/64	49/64	1/16
R-2404	Fits Am. Std. OEM 305-17, Kohler	55/64	41/64	3/32
R-2405	Fits Am. Std., Central, Eljer, Harcraft, Kohler, Speakman	3/4	29/64	5/64
R-2405A	Fits Kohler, Speakman	25/32	29/64	5/64
R-2406	Fits Am. Std. OEM 54109, Crane, Crosley, Kohler, Price-Pfister, Repcal, Schaible, Tracy, Youngstown	7/8	29/64	5/64
R-2406A	Fits Speakman	29/32	29/64	5/64
R-2407	Fits Am. Std. OEM 681-17, Repcal	1 13/64	1 5/64	3/64
R-2408	Fits Am. Std. OEM 670-17	3/4	19/32	3/64
R-2409	Fits Am. Std., Barnes, Briggs. Milwaukee, Price-Pfister, Sterling	55/64	29/64	5/64
R-2409A	Fits Barnes, Union Brass	13/16	25/64	3/32
R-2409B	Fits Am. Kitchen, Gerber, Schaible	55/64	25/64	5/64

FRICTION RINGS

PACKING RINGS FOR SWING SPOUT, SLIP-JOINTS, PUMP AND HYDRAULIC APPLICATIONS

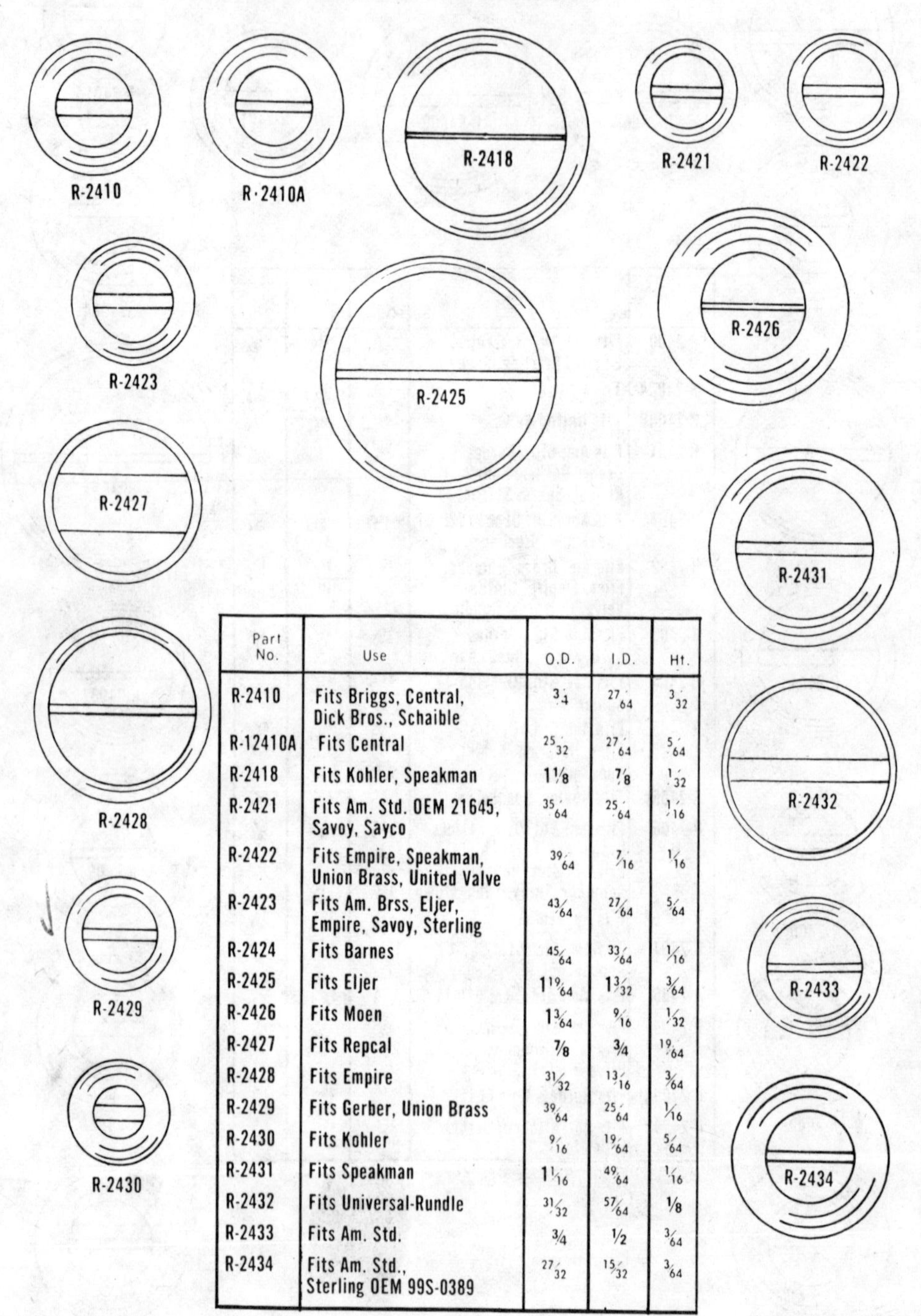

Part No.	Use	O.D.	I.D.	Ht.
R-2410	Fits Briggs, Central, Dick Bros., Schaible	3/4	27/64	3/32
R-12410A	Fits Central	25/32	27/64	5/64
R-2418	Fits Kohler, Speakman	1 1/8	7/8	1/32
R-2421	Fits Am. Std. OEM 21645, Savoy, Sayco	35/64	25/64	1/16
R-2422	Fits Empire, Speakman, Union Brass, United Valve	39/64	7/16	1/16
R-2423	Fits Am. Brss, Eljer, Empire, Savoy, Sterling	43/64	27/64	5/64
R-2424	Fits Barnes	45/64	33/64	1/16
R-2425	Fits Eljer	1 19/64	1 3/32	3/64
R-2426	Fits Moen	1 3/64	9/16	1/32
R-2427	Fits Repcal	7/8	3/4	19/64
R-2428	Fits Empire	31/32	13/16	3/64
R-2429	Fits Gerber, Union Brass	39/64	25/64	1/16
R-2430	Fits Kohler	9/16	19/64	5/64
R-2431	Fits Speakman	1 1/16	49/64	1/16
R-2432	Fits Universal-Rundle	31/32	57/64	1/8
R-2433	Fits Am. Std.	3/4	1/2	3/64
R-2434	Fits Am. Std., Sterling OEM 99S-0389	27/32	15/32	3/64

BEVELED (CONE) SLIP JOINT WASHERS

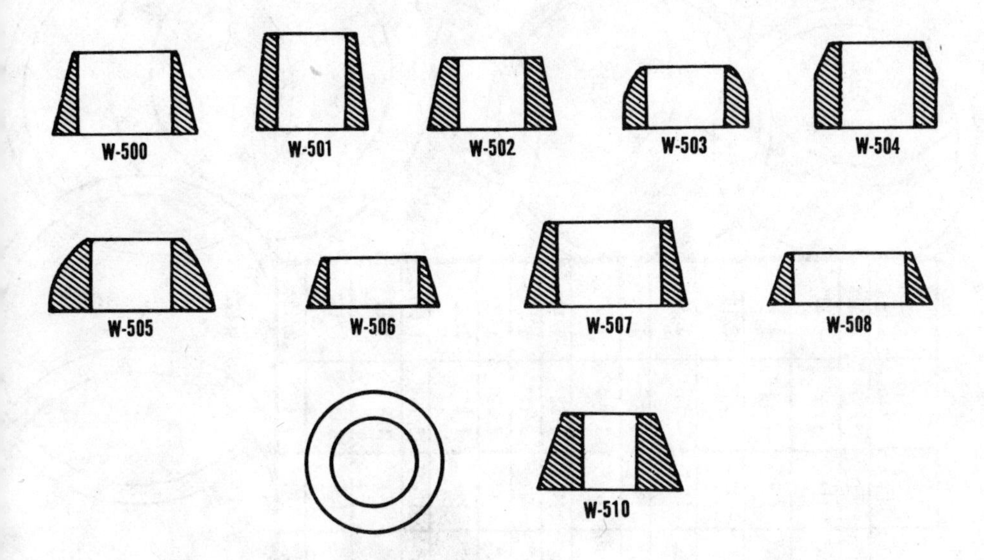

Part No.	Hole	O.D.	Height	Purpose	Use Brass Friction Ring No.
LAVATORY SUPPLY					
W-501	3/8″	5/8″	1/2″	3/8″ O.D. Tube, 1/2″ Comp. Nut Threaded Speedy Connectors	R-503
W-502	3/8″	23/32″	3/8″	3/8″ O.D. Tube, 1/2″ I.P. Nut	R-507
W-503	7/16″	5/8″	5/16″	1/2″ O.D. Tube, 1/2″ Comp. Nut	R-501
W-506	1/2″	23/32″	1/4″	1/2″ O.D. Tube, 1/2″ I.P. Nut	R-502
CLOSET SUPPLY					
W-510	5/16″	27/32″	3/8″	3/8″ O.D. Tube, Ballcock Shank Fits Reamed & Not Reamed	R-509
W-500	1/2″	27/32″	7/16″	1/2″ O.D. Tube, Ballcock Shank	R-506
W-505	5/8″	27/32″	9/32″	3/8″ I.P. Ballcock Shank 1/2″ Copper Water Tube	R-508
W-508	7/16″	7/8″	3/8″	1/2″ O.D. Tube for Old Kohler	R-506
BATH SUPPLY					
W-505	7/16″	7/8″	3/8″	1/2″ O.D. Tube, 1/2″ I.P. Nut Botttom of Supply	R-506
W-507	9/16″	7/8″	7/16″	9/16″ O.D. Tube, Top of Supply	R-507
W-500	1/2″	27/32″	7/16″	9/16″ O.D. Tube, Botttom of Supply	—
WEST COAST LAVATORY SUPPLY					
W-501	3/8″	5/8″	1/2″	7/16″ O.D. Tube, 1/2″ Comp. Nut—Long Taper	R-503
W-502	3/8″	23/32″	3/8″	7/16″ O.D. Tube, 1/2″ I.P. Nut	R-501
W-504	7/16″	5/8″	7/16″	7/16″ O.D. Tube, 1/2″ I.P. Nut	R-501
W-506A	7/16″	23/32″	1/4″	7/16″ O.D. Tube, 1/4″ I.P. Nut Nut—Long Taper	R-506
WEST COAST CLOSET SUPPLY					
W-505	7/16″	7/8″	3/8″	7/16″ O.D. Tube, Ballcock Shank	R-506

SQUARE CUT SLIP-JOINT WASHERS

Part No.	Trade Size	Enters Nut	To Slip Over	±1/64" O.D.	±1/64" I.D.	O.D.-I.D. Wall	±1/32" Height
W-511	27	Supply	3/8" Tube	1/2"	3/8"	1/16"	7/32"
W-515	2A Thin	1 1/4" I.P.	1 1/4" Tube	1 25/64"	1 1/4"	1/16"	3/16"
W-517	6A Thin	1 1/2" I.P.	1 1/2" Tube	1 41/64"	1 1/2"	1/16"	3/16"
W-518	7A Thin	2" I.P.	2" Tube	2 9/64"	2"	1/16"	3/16"
W-519	15 Thin	1 1/4" I.P.	1 1/4" Tube	1 13/32"	1 7/32"	3/32"	3/32"
W-520	2	1 1/4" I.P.	1 1/4" Tube	1 7/16"	1 1/4"	3/32"	3/16"
W-521	12 Thick	1 1/2" I.P.	1 1/2" Tube	1 11/16"	1 7/16"	1/8"	1/8"
W-522	6	1 1/2" I.P.	1 1/2" Tube	1 11/16"	1 1/2"	3/32"	3/16"
W-523	7	2" I.P.	2" Tube	2 3/16"	2"	3/32"	3/16"
W-530	1	1" I.P.	1" Tube	1 3/16"	1"	3/32"	3/16"
W-531	2B Thick	1 1/4" I.P.	1 1/4" Tube	1 1/2"	1 1/4"	1/8"	3/16"
W-532	14	1 1/2"x2"	1 1/4" Tube	2 1/8"	1 1/2"	5/16"	1/8"
W-533	6B Thick	1 1/2" I.P.	1 1/2" Tube	1 3/4"	1 1/2"	1/8"	3/16"
W-534	7B Thick	2" I.P.	2" Tube	2 1/8"	1 7/8"	1/8"	3/16"
W-535	3	1 1/2" I.P.	1 1/4" Tube	1 23/32"	1 7/32"	1/4"	3/16"

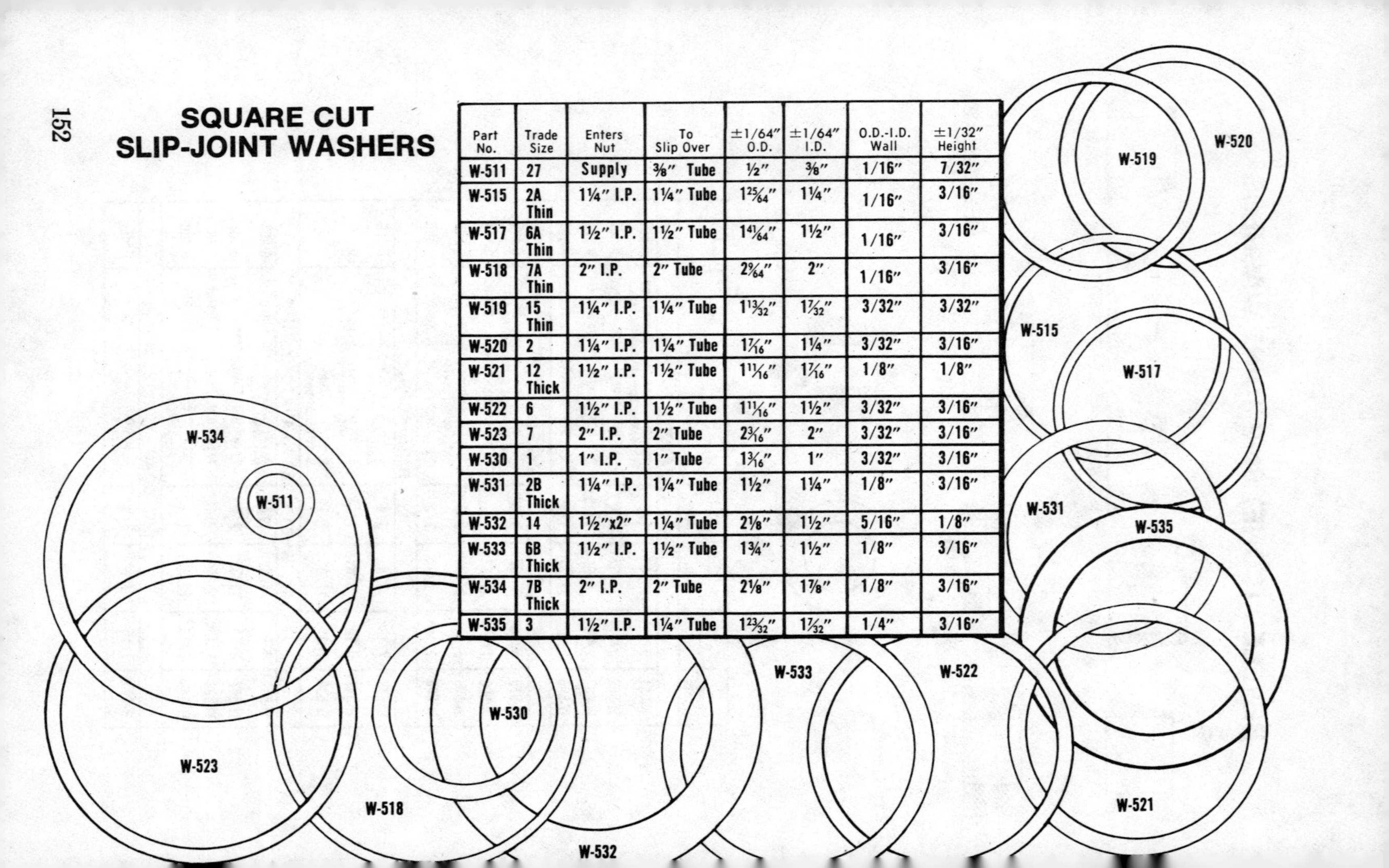

GLOSSARY OF WORDS USED IN PLUMBING

FSPS Female standard pipe size.
MPT Male pipe thread.
MSPS Male standard pipe size.

ADAPTER A fitting used to connect two different size pipe or fitting. For Example: Soil Pipe Adapter joins cast iron soil pipe hub to 3", 2" or 1-1/2" copper tube. Available in many different sizes in both male and female.

BACKING BOARD Usually a 1x4, 1x6 or 1x8 mortised in and nailed flush with leading edge of studs to provide support for a fixture.

BIDET (pronounced be-day) A fixture that cleanses the essentials.

BRANCH Any vent or drainage pipe other than soil or vent stack.

BRIDGING A support or nailor nailed between joists or studs.

CALKING Also spelled caulking. Material like oakum and lead.

CALKING TOOLS iron, plumbers furnace, lead pot, ladle.

CAST IRON Hub and Spigot.

CAST IRON CALKING RUNNER Asbestos rope clamp. Permits pouring lead in horizontal hub and spigot joint.

CAST IRON PLUMBING TREE

CAST IRON PIPE CUTTER

CAT Short piece of lumber, usually 2x4, nailed between studs to back up edge of gypsum board or fixture.

CHASE Recess cut in framing to permit installing pipe.

CLEANOUT PLUG cast iron.

CLOSET BEND

CLOSET FLANGE Also called floor flange.

CLOSET SCREW Long screw with detachable head formerly in wide use for fastening water closet to floor.

CORNER TOILET

COUPLING A coupling joins two pieces of pipe of the same or different sizes. Some couplings have stops to allow pipe to only go in so far; others have no stop. Also available Copper to Slip-Joint.

DIVERTER VALVE bathtub and shower control. Connects to hot and cold water line.

DRAINAGE Any pipe that carries waste water in the drainage system.

DROP EAR ELBOW 90° 1/2" x 3/4". Permits connecting copper shower supply line to threaded nipple required for shower head.

DRUM TRAP recommended for bathtub installation.

DWV FITTINGS these drainage, waste and vent fittings incorporate the recommended drainage pitch of 1/4" to the foot.

ELL — L — ELBOW a Quarter Bend 90°.

ESCUTCHEON A plate used to enclose pipe or fitting at wall or floor opening.

FEMALE end of fitting receives male.

FERRULE a threaded sleeve soldered to hub of pipe.

FITTINGS Any coupling, tee, elbow, union, etc., other than pipe. Plumbing catalogs refer to fittings as "ftg."

FIXTURE PLACEMENT CHART p. 73.

FIXTURE UNIT A method of estimating amount of water a fixture discharges. A unit is equivalent to 7-1/2 gallons of water or one cubic foot of water per minute. While a bathroom containing toilet, lavatory, bathtub or shower stall is rated by national codes as 6 units, the same codes rate a bathtub with 1-1/2" trap, with or without shower, 2 units; with 2" trap — 3 units; a bidet — 3 units; lavatory — 1 unit; shower stall — 2 units; an extra toilet with 3" drain — 4 units.

FPT Indicates female pipe thread.

FRAMING walls for bathroom.

FRESH AIR INLET Pipe above roof. Codes frequently require this be size larger than internal vent line.

INCREASER A coupling with one end larger than the other. Used to increase diameter of pipe above roof.

INSPECTION PANEL provides access to bathtub trap.

KAYFER Also called Kafir. A screw type hub fitting on cast iron that simplifies making new connection in 4" soil line.

LAVATORY P TRAP

LEAD BEND formerly used exclusively as a closet connection to soil line.

MALE End of fitting inserts in female.

NO-HUB CAST IRON

OFFSET Any combination of pipe and fitting, or combination of fittings used to angle over.

PARTITION end of tub.

PIPE CUTTER

PIPE MARKER insures sawing pipe square.

PIPE SIZING CHART Simplifies sizing existing pipes.

PIPE STRAP

PLUMBING TOOLS

PRE-ENGINEERED PLUMBING WALL a completely assembled plumbing wall.

PRE-FABRICATED BATHROOM see Book #682.

REDUCER Copper to copper. Joins 3/4" to 1/2". Also available in other sizes.

ROOF VENT INCREASER Fitting simplifies increasing vent from 3" to 4". Available 3" x 4" in 18", 24", 30" lengths.

ROUGH—IN This describes installation of drainage waste and supply lines. Roughing-in concerns all work required prior to connecting fixtures.

SANITARY TEE 90° WITH SLIP JOINT Joins copper waste and vent line to chrome or brass lavatory or sink drain pipe. Nut tightens lead ring to make tight joint.

SLIP JOINT

SOIL STACK Codes allow 3" or 4" cast iron, copper, plastic.

SOIL LINE LAYOUT GUIDE use folding rule or garden hose, see Book #682.

SOIL PIPE carries discharge from one or more toilets and/or discharge from other fixtures to main sewer line.

SOLDER CUP END

SPIGOT end of pipe that fits into hub.

STACK VENT OR VENT STACK that part of the soil stack above the highest drain connected to stack.

STOP, VALVE Available in angle and straight stops. Permits shutting off supply to fixture when repairs are required.

TOILET roughing-in.

TRAP A fitting designed to provide a liquid seal to prevent back passage of air. P-Trap, Drum Trap, House Trap.

TUBE BENDER tool simplifies bending soft copper tubing.

TUBE STRAP

VENT STACK this is a vertical pipe that provides circulation of air to branch vents, revents or individual vents.

WALL HUNG TOILET

WASTE this refers to water from any fixture except toilet.

WASTE PIPE one that conveys only liquids, no fecal matter.

WET VENT a wet vent is both a vent and drainage line from any fixture except a toilet.

Y-BRANCH

Plumbers, like most skilled craftsmen, and other self employed —doctors, dentists and lawyers, frequently charge whatever they believe the customer can afford. Adjusting a faulty toilet, opening a stopped up drain, drilling and filling a tooth, or prescribing a remedy, have one common denominator, each requires the sale of time and experience. Each segment of society earns its living rendering service to others. What you pay depends entirely on your knowledge and willingness to do the work yourself. Keep well and you save doctors bills. Keep plumbing in repair and you effect comparable savings.

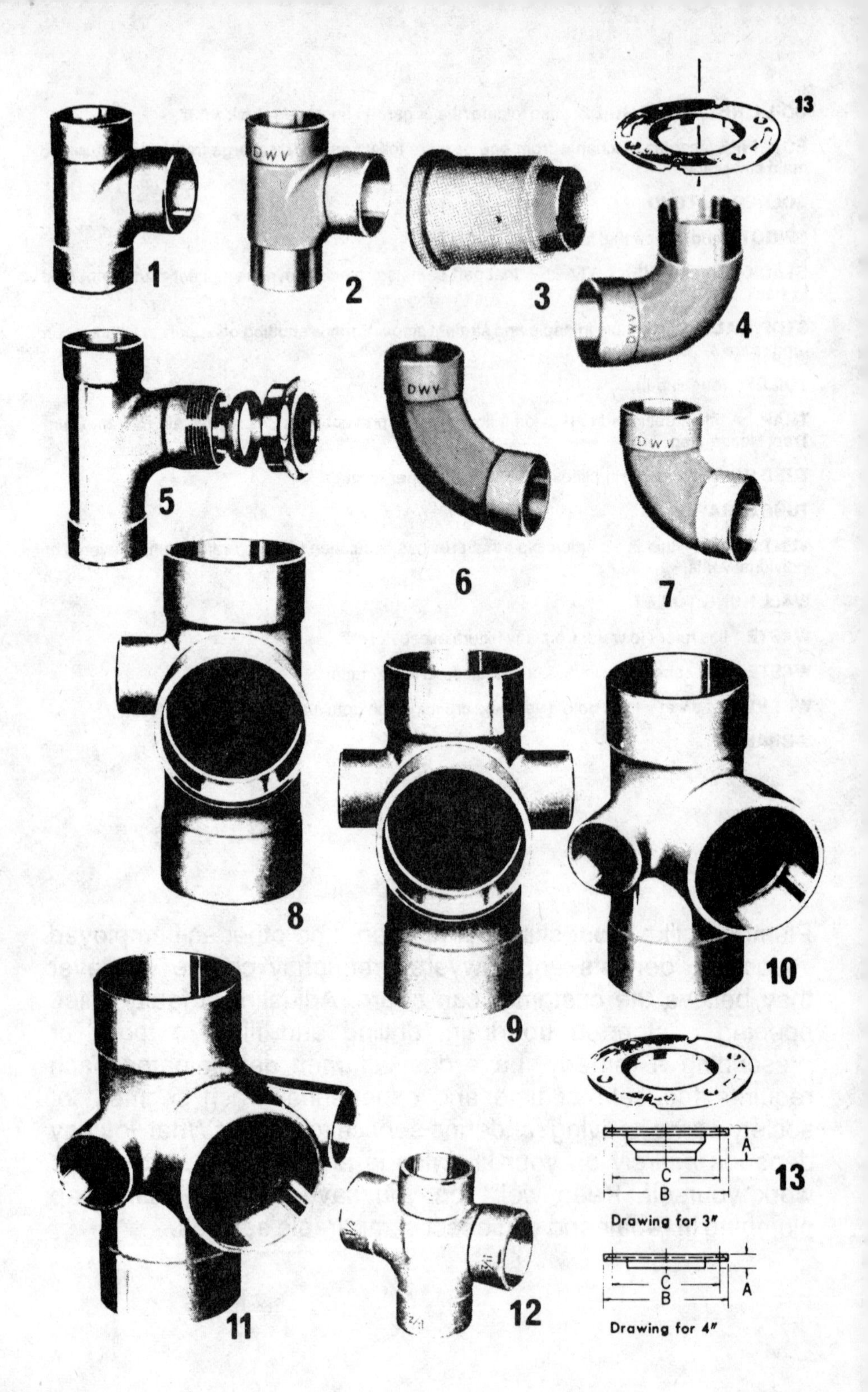
13
1
2
3
4
5
6
7
8
9
10
11
12
13
Drawing for 3"
Drawing for 4"

COPPER FITTINGS

1 Sanitary Tee. Copper to Copper to Copper.
Available in many different sizes. To indicate size specify number 1, 2 and 3 in order indicated.

2 Fitting Sanitary Tee. Fitting to Copper to Copper.

3 Soil Pipe Adapter. Connects Soil Pipe Hub to Copper tube.
Available in many different sizes.

4 Quarter Bend 90°. Copper to Fitting.
Also used as a closet bend with floor flange No. 13.

5 90° Sanitary Tee with Slip-Joint.
Joins copper tube to chrome or brass drain from lavatory or bathtub.

6 Quarter Bend 90°. Copper to Copper long turn.
Joins two lengths of copper tube.

7 Quarter Bend 90°. Copper to Copper.

8 Sanitary Tee with Side Inlet on Left.
Copper to Copper to Copper to Copper.

9 Sanitary Tee with Side Inlet on Left and Right.

10 Stack Fitting with Left Inlet.
Provides inlet for toilet, also 1½″ or 2″ inlet for drainage line.

11 Stack Fitting with Right and Left Inlet.

12 Double Sanitary Tee. Copper to Copper to Copper to Copper.
Used where code requires separate vents to main stack.
Available 3″ x 3″ x 1½″ x 1½″ and many other sizes.

13 Closet Flange attaches toilet to closet bend. Quarter Bend 90°, Illus. 4, can be used as a closet bend.

COPPER FITTINGS

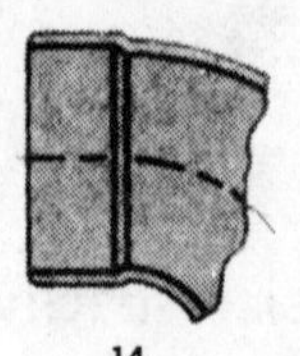

14
Solder Cup End.

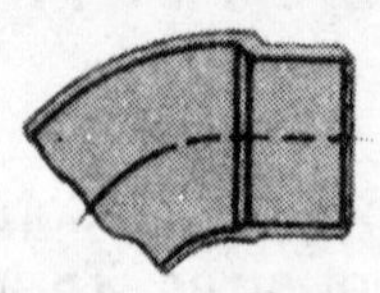

15
Fitting End. This end goes into fitting.

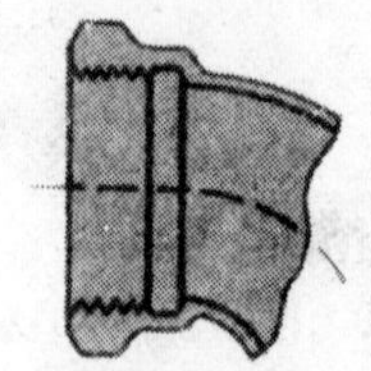

16
Usually designated FPT or FSPS. Female pipe thread or female standard pipe size.

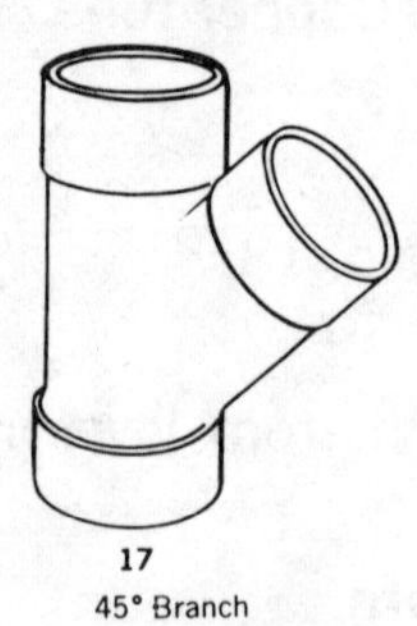

17
45° Branch
Joins branch line to main line.

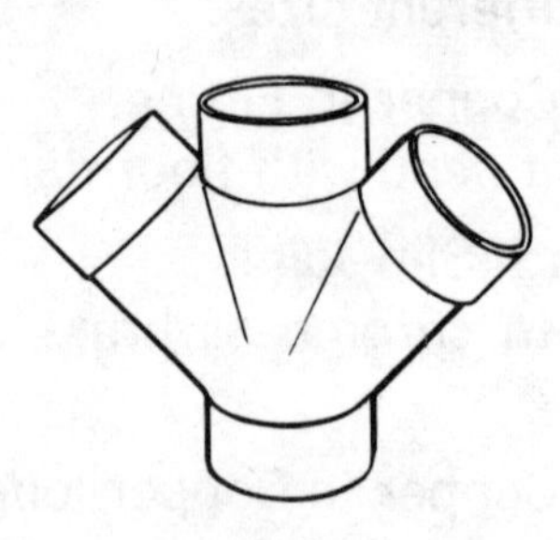

18
45° Double Y-Branch.

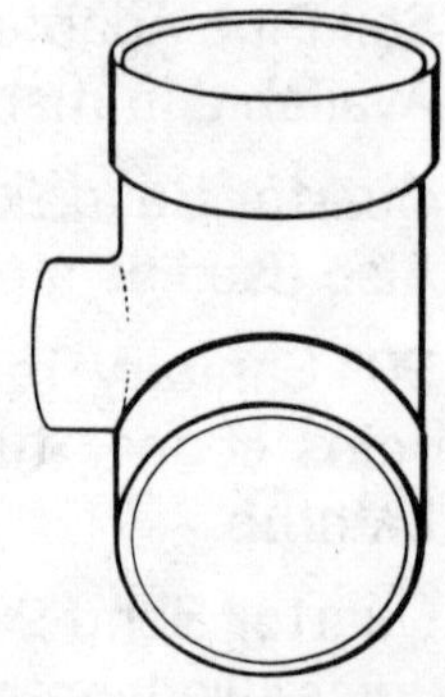

19
Side inlet serves a lavatory or bathtub.

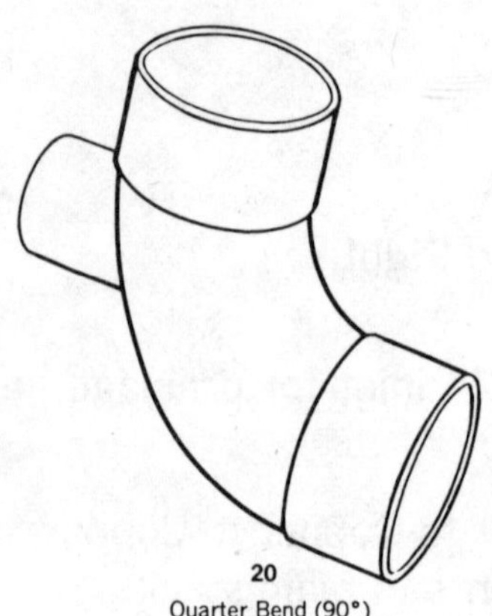

20
Quarter Bend (90°)
High Heel Inlet.

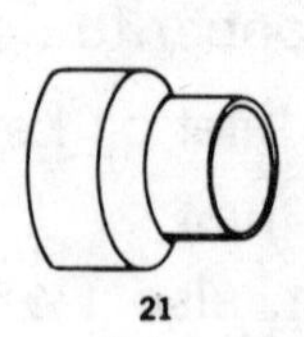

21
Reducer

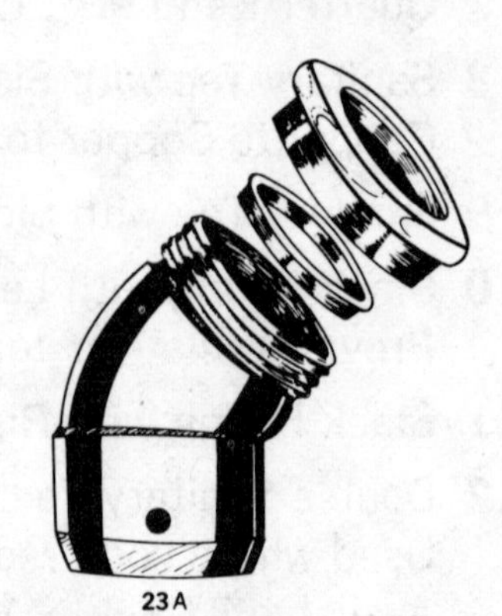

23 A
Drainage (45°) Slip Joint Adapter

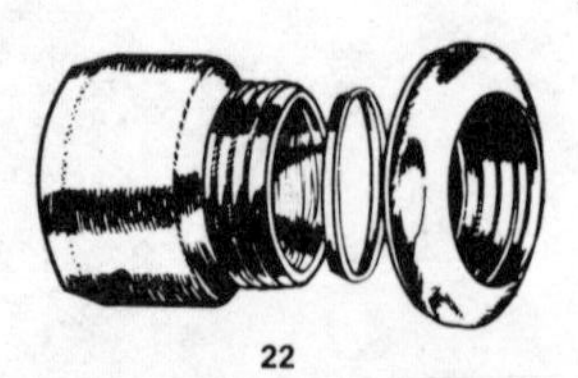

22
Drainage Straight Slip Joint Adapter

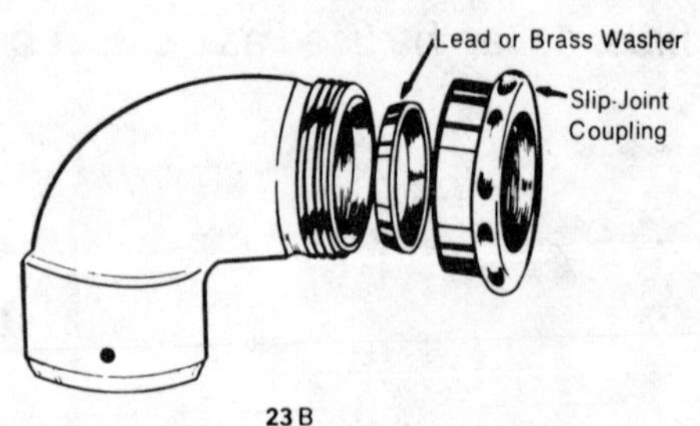

23 B
Drainage (90°) Slip Joint Adapter

COPPER FITTINGS

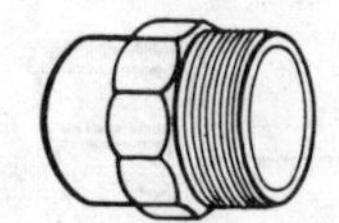

24
Male Adapter — copper tube to threaded female.

25
Male Fitting Adapter — Joins fitting to threaded female.

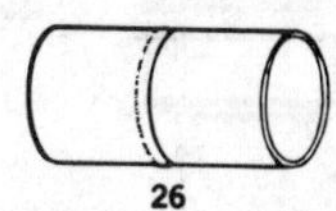

26
Coupling with stop.

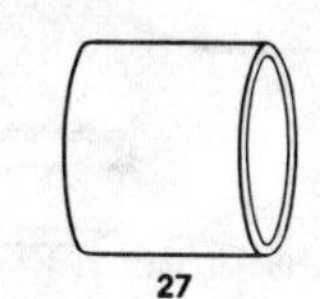

27
Repair coupling. No stop.

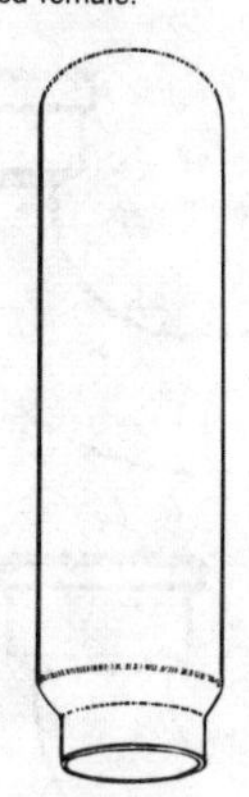

28
Roof Vent Increaser — 3″ x 4″ x 18″ or 24″ or 30″ long.

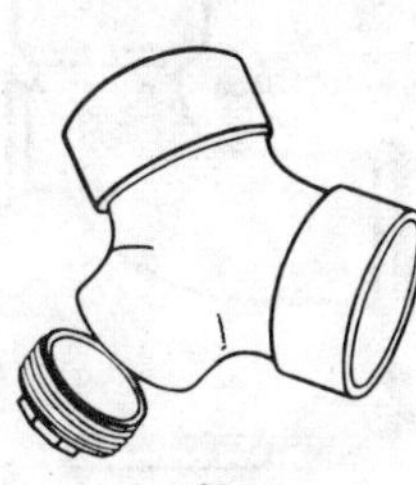

29
Eighth Bend (45°) with cleanout.

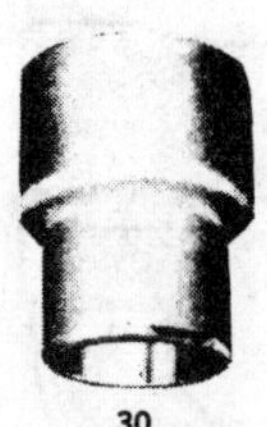

30
Vent Increaser — Copper to Copper.
Can be used to change diameter of vent stack just before going through roof. Available in wide combination of sizes — 1½″ x 4″, 1½″ x 3″, etc.

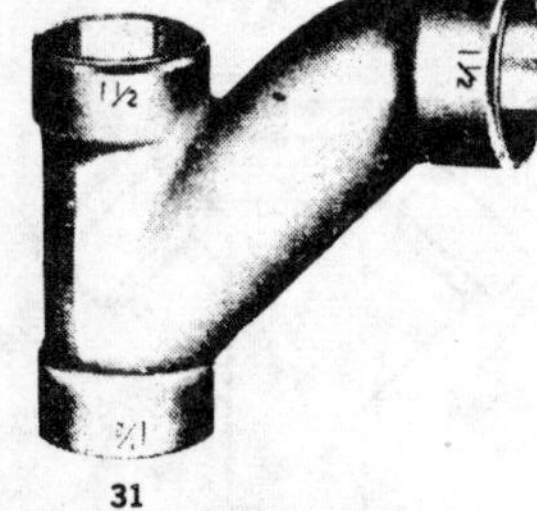

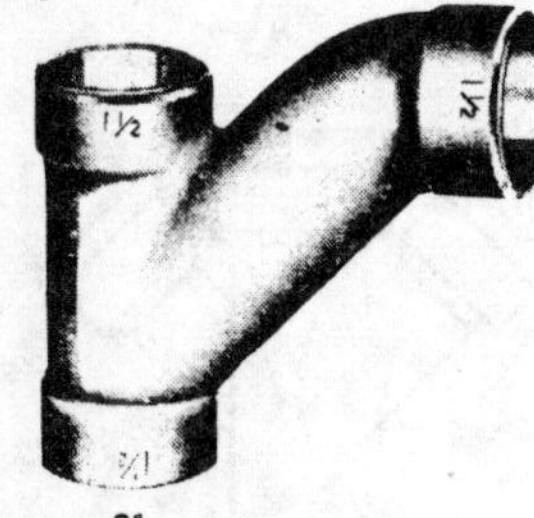

31
Long Turn T-Y. Also available with side inlet on branch.

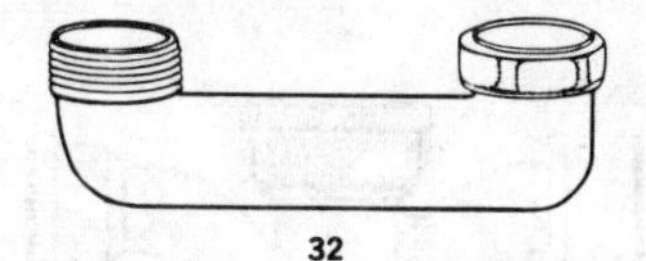

32
Upturn fastens to bottom of one piece drum trap
Bathtub outlet connects to Slip Joint.

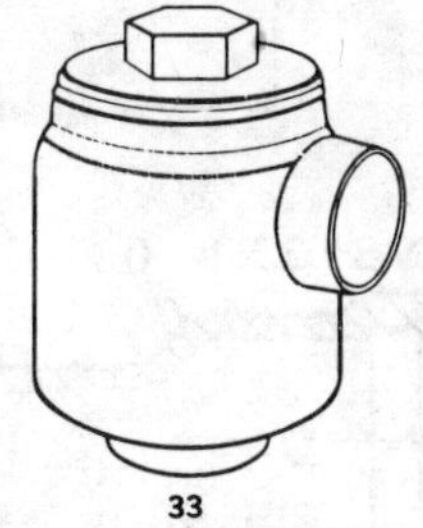

33
Drum Trap — Inlet from bathtub in bottom, outlet in top.

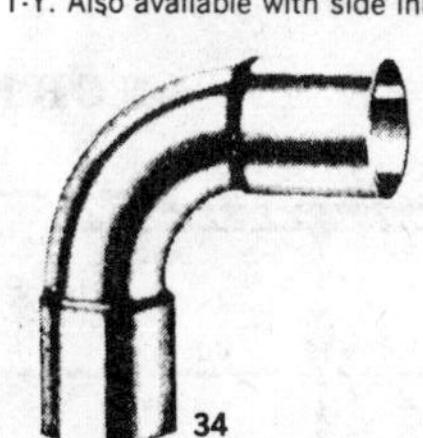

34
90° Elbow — Copper to Copper.
Available in short or long radius.

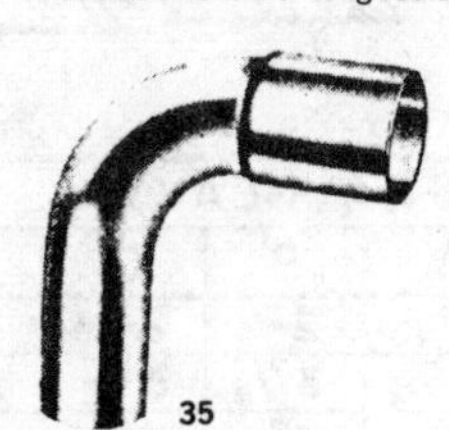

35
90° Ell Short Radius, also available long radius.
Copper to Fitting.

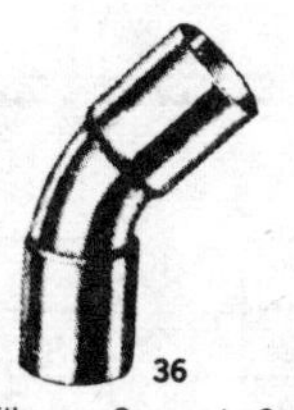

36
45° Elbow — Copper to Copper.

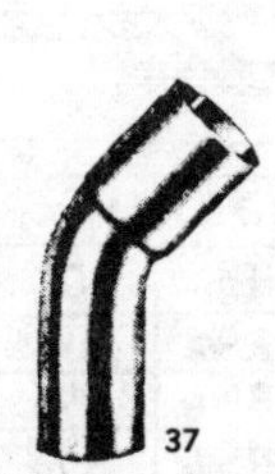

37
45° Fitting to copper ell.

38 Cap.

CAST IRON

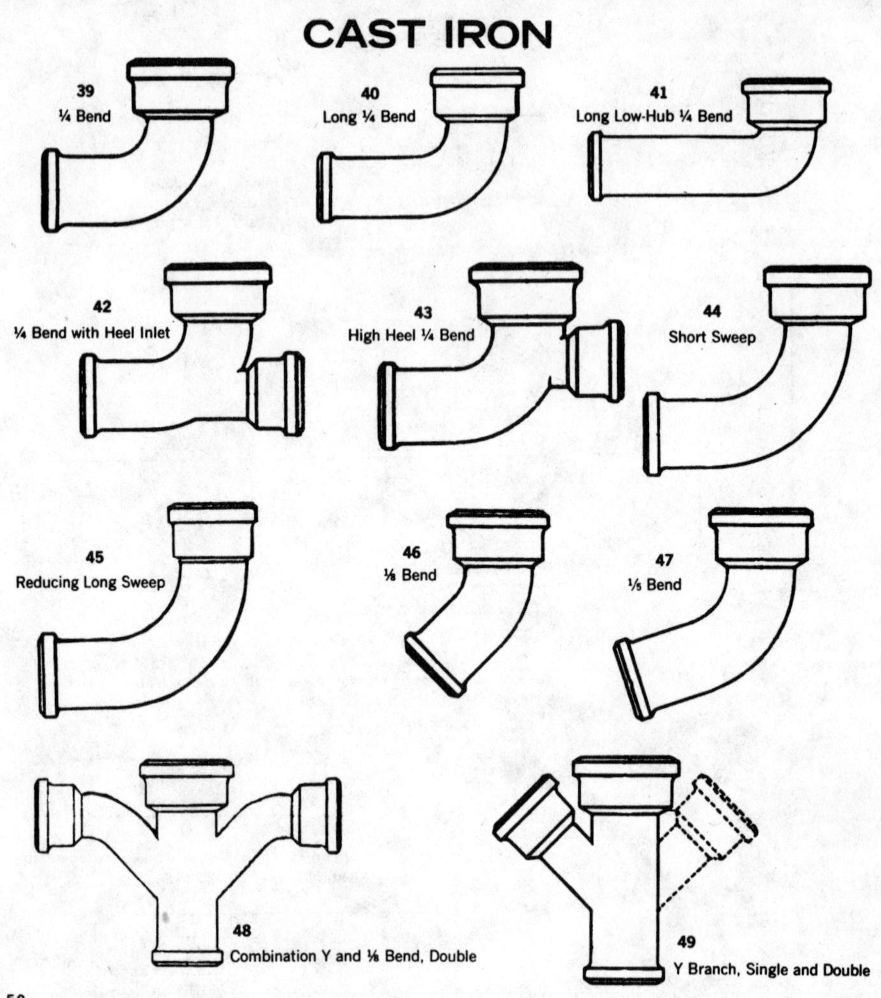

50
DIMENSIONS OF CAST IRON SOIL PIPE IN INCHES

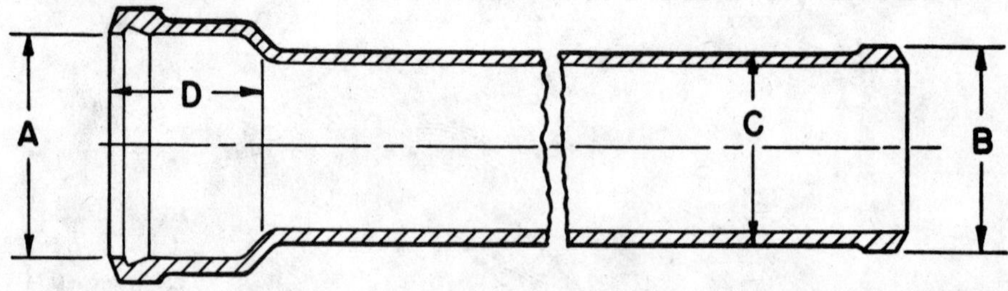

STANDARD				
SIZE	A	B	C	D
2	2 15/16	2 5/8	2 1/4	2 1/2
3	3 15/16	3 5/8	3 1/4	2 3/4
4	4 15/16	4 5/8	4 1/4	3

EXTRA HEAVY				
SIZE	A	B	C	D
2	3 1/16	2 3/4	2 3/8	2 1/2
3	4 3/16	3 7/8	3 1/2	2 3/4
4	5 3/16	4 7/8	4 1/2	3